A New Basis for Animal Ethics

A New Basis for Animal Ethics

Telos and Common Sense

Bernard E. Rollin

UNIVERSITY OF MISSOURI PRESS
Columbia

ISBN: 978-0-8262-2101-8
Library of Congress Control Number: 2016944654

∞™ This paper meets the requirements of the
American National Standard for Permanence of Paper
for Printed Library Materials, Z39.48, 1984.

Typefaces: Minion, Myriad

This book is dedicated to the memory of my grandmother, Anna Bookchin, who instilled in me the toughness necessary to fight for the defenseless, and to my mother, who taught me empathy and compassion for all living things.

And also to my family: My wife, Linda, perennial wise critic and partner in dialogue; my son, Michael, endless source of inspiration and fountain of ideas; my daughter-in-law, Theresa; and my grandchildren, Danny and Lily.

Contents

Preface

FOR THE LAST forty-five years, I have worked to raise the moral status of animals in society, on both a theoretical and a practical level. Unlike many philosophers, I have been able to effect significant change in animal use in society, to the direct benefit of the animals. Other, like-minded people and I have made major changes in the ways animals are used in education, eliminating many of the atrocious exercises that were earlier seen as essential to becoming a veterinarian or a physician or a science professor. We have been able to establish the control of pain in research as a major duty of the responsible researcher, and we have encoded this duty in legislation. We have been able to catalyze the elimination of one of the most egregiously inhumane housing systems regnant in confinement agriculture: sow stalls, or gestation crates. Equally important, we have been able to occasion moral thinking regarding our ethical obligations to animals among citizens and animal users alike.

This book is an account of the thinking underlying these far-reaching changes. It is based in the realization that one's ethics cannot be separated from one's worldview or metaphysics. A metaphysic that sees the world as simply made of material particles obeying the mechanistic laws of physics, as, for example, Descartes postulated, is going to see such a world inevitably as having no place for values, and particularly, no place for ethics. Such is the world envisioned by physics developed during and after the Renaissance. In such a world it is understandable how scientists could affirm that a scientific description of the world is "value-free" and leaves no conceptual room for ethics.

Happily, this is not the only way of looking at the world. Ordinary common sense sees the world as replete with qualitative differences not capturable by the language of mathematical physics alone. In the world of our experience we find beautiful and ugly, living and nonliving, good and bad, right and wrong, and the entire vast array of qualities that make the world an exciting and challenging place to live. The metaphysics of that world was best captured by Aristotle, particularly in his emphasis on telos as the core explanatory concept for the world we live in. We understand what an animal is by what it *does*—the "pigness" of the pig, the "dogness" of the dog. This is the biology that ordinary people understand, the study of living organisms as they live, not as reduced to their molecular components. This is the *nature* of an animal, and considering this helps us understand our obligations to animals even as an understanding of human nature helps determine our obligation to humans. In this book, then, I will explicate and justify our moral obligation to animals in terms of the commonsense metaphysics and ethics of telos. By so doing, I hope to introduce ordinary, commonsense people to a set of obligations to animals following from their own beliefs. I hope to, as Plato said, lead people to *recollect* their own ethic, not teach them something new.

It took decades, but I can finally see a unity in the disparate ideas I have developed over my career.

Acknowledgments

IN A REAL sense, this book is a product of more than 35 years of thinking, writing, teaching, lecturing and acting in the area of animal ethics. To acknowledge everyone to whom I owe significant gratitude for dialogue and criticism over the years would itself require a book. Nonetheless, there are certain people to whom I owe a recent debt. These people include my wife, Linda; David Rosenbaum, Director of the University of Missouri Press, who has been a dear friend and brilliant critic who helped me in innumerable ways for over 20 years; the readers and editors for the University of Missouri Press, Gloria Thomas and Sara Davis; and my friend and colleague and co-teacher Terry Engle. And I must also thank my students at Colorado State University and lecture audiences all over the world, who help assure that I know what I am talking about.

A New Basis for Animal Ethics

Introduction
Philosophy and Ethics

IT IS QUITE patent that college students are often drawn to studying philosophy, or indeed, majoring in philosophy, by something one can call the "Wow! factor." How many eighteen-year-olds can fail to be charmed and enchanted by Zeno's proof that motion is impossible? Or by McTaggart's demonstration that time is unreal? Or by Hempel's paradox showing that a piece of white chalk serves as evidence for the claim that all ravens are black? All such dazzling arguments elicit a "Wow!" when they are first encountered. For students who have lived a sheltered life, Hume's attack on the religious Argument from Design that purports to show the existence of an intelligent designer can be similarly mind expanding.

College students tend to be attracted either to intellectual approaches that radically challenge their understanding of reality, such as the paradoxes mentioned above, or else to those at the other extreme that promise to fix and reform the world, such as Marxism. A famous quip from Bertrand Russell well characterizes the latter tendency: "If you are not a radical at age twenty, you are a knave. And if you are a radical at age forty, you are a fool." Twentieth-century Anglo-American analytic philosophy was adamant about the role of philosophy in real-world matters. Ludwig Wittgenstein, the patron saint of analytic and ordinary-language philosophy, set the tone for analytic approaches to ethics when he remarked that one can take an inventory of all the facts in the world and not find it a fact that "killing is wrong." Thus mainstream British and American philosophers did not embrace real-world issues, instead engaging in "meta-ethics," thinking about the nature of ethical judgments. The dominant approach to ethics was "emotivism," which claimed that apparent ethical judgments that seem to be making claims about reality are really just expressions of emotions such as revulsion or disgust. Claiming that killing is wrong is to be analyzed as analogous to saying, "Killing, yuck!" Thus one cannot rationally argue about ethics. When we appear to argue about an ethical issue, such as the illegitimacy of capital punishment, we are really arguing about such *facts* as whether capital punishment deters other killers. Needless to say, such an approach to ethics is very unlikely to have the Wow! factor for students seeking to change the world, or even to generate intellectual excitement for them.

In my own case, both in my undergraduate studies and in graduate school, I could develop no enthusiasm for ethics. For a smart young person, the meta-ethics that was regnant during my schooling came as no surprise— *of course* ethical terms do not refer to, nor can they be explained by, natural, empirical properties. And the contrary notion, that ethical terms refer to abstract entities that "subsist" in a Platonic realm of Forms, was at least equally implausible. Only much later in my career could I see the power of arguments purporting to lead to the postulating of abstract entities. And it became clear to me that various theories intended to explain our ethical intuitions as deployed in daily life lack plausibility. Regarding utilitarianism, which defines the ethical notions of "good" and "bad" in terms of those actions that result in the greatest amount of pleasure for the greatest number and the least amount of pain, it is difficult to see any way of adding up pleasures and pains in any consistent way. For example, how does one compare physical pain with psychological distress in a way that lends itself to a felicific calculus that can be consistently applied? How, for example, does one weigh the pain of a back injury in football against the risks involved in continuing to play or the loss of a lucrative potential career? Obvious problems arose as well with Kantian ethics—how could anyone buy into the idea that there may never have been a moral act or the related idea that if one derives even an iota of pleasure from an action, it ceases to be moral? According to Immanuel Kant, for an action to have any moral worth, it must be performed strictly out of acknowledgment and respect for the moral law. Thus, for example, if one is contributing to charity, say, buying Christmas presents for destitute children, and if one consequently derives pleasure or satisfaction from contemplating the children's joy, the action cannot be viewed as strictly moral, for it is possible that the satisfaction one derives is the major reason for the action.

In short, ethics did not elicit a "Wow!" from me. In part, this is because we all make mundane ethical judgments all the time. Very rarely is an ethical decision highly dramatic the way ethical problems are depicted in the media—two patients, only enough medicine for one, who lives and who dies? Even in medicine, such decisions are extremely rare. ("Thank God!" say my physician friends.) When genuine dilemmas do arise, rarely is one forced into a philosophical examination of one's fundamental assumptions, in the way that Zeno's paradoxes do drive such conceptual reflection about space, time, and motion. Rather, the agent will attempt to find some morally relevant characteristic that sways the resolution one way or the other. (In the medical example above, one patient may be a mass murderer of children, a fact that would help break the deadlock over who receives the medicine.)

Thus, for many young philosophers trained in the 1950s, 1960s, and 1970s, ethics seemed to be rather dull, as it failed to produce shocking conclusions and was specifically taught as having nothing to do with changing the world. While some hushed talk about "applied ethics" could be heard by the 1970s, it was spoken of derisively and never addressed in the "good" philosophy schools. All of this was quite odd, given the social turmoil extant during this period. When I began to work on animal ethics in the mid-1970s, I received a number of letters from former classmates at Columbia University, warning me not to get sidetracked from "real philosophy." I must confess to partaking in this snobby and snotty attitude myself for quite a long time. And with the chutzpah shared by the very bright and the very dumb, analytic philosophers dismissed the history of philosophy as irrelevant to a sound philosophical education because, after all, it is a history of errors. That rankled, both because it was my specialty and because even Aristotle, arguably one of the two greatest philosophers in history, began all his inquiries by examining what his predecessors had said about the subject in question, and *he had virtually no predecessors.*

Part One

Creating an Animal Ethic

The Need for a New Animal Ethic

ETHICS BECAME FAR more interesting to me in the mid-1970s by virtue of a unique congeries of circumstances. First of all, in the course of teaching the history of philosophy seven hours per week each year for eight years, I enjoyed a synoptic view of the issues philosophers were preoccupied with, including ethical issues. Second, by virtue of paying close attention to the news, it became clear to me that the social-ethical "searchlight" was beginning to focus on animal matters (as well as environmental concerns), yet there was no viable ethic to direct, sustain, and guide progress in this area (environmental ethics also began to develop then). Third, I had a longstanding interest in animals and some awareness that they were not receiving the best possible treatment in accordance with their societal uses. By virtue of this interest, I began to look at the history of philosophy from a different perspective, specifically searching for reasons philosophers gave to exclude animals from full moral attention and consideration, and I found their arguments grossly inadequate. Fourth, in the course of reexamining the history of philosophy, I found very little discussion of the moral status of animals, with two notable exceptions. Fifth, I had been approached by a key faculty member in the College of Veterinary Medicine at Colorado State University, where I teach, about creating a course in veterinary medical ethics, focusing on the changing societal views of animals and what practical implications these had for veterinary medicine. All of this led me to try to write something helpful on the moral status of animals, which in turn made me begin to understand how little protection animals enjoyed in all areas of social use and how desperately additional mechanisms for animal protection were needed.

All of this also led me for the first time to engage in an authentic way the philosophical issue of how one develops a new ethic for anything and, equally importantly, how one persuades others to look sympathetically at the results of such thinking. I found myself greatly vexed and perplexed by these questions, and falling rapidly and headlong into genuine philosophical thinking about ethics that would shape my activities, both theoretical and practical, for the next four decades. In the course of my thinking, it became clear that there existed strong constraints on the development of a new ethics for animals:

1. Any putative ethic proposed to society must be both difficult for citizens to reject and easy for them to accept, at least on a theoretical level. As we will discuss in detail shortly, in Plato's terms one must aim at *reminding* rather than *teaching*. In other words, the suggested ethic must resonate with people's already deeply held beliefs. This strategy was successfully deployed by Martin Luther King Jr. and Lyndon Johnson regarding civil rights and segregation. If the ethic suggested for animals affirms or entails that animals should have the right to vote, to take an absurd example, that will have no traction with the general public.

2. Those promulgating the new ethic must not seek to establish it too quickly. The great radical activist Henry Spira often remarked that all social-ethical revolutions in the history of the United States have been gradual. To expect people to suddenly abandon established cherished practices is impracticable and unrealistic.

3. One should seek a middle ground between extremes. For example, in the case of invasive experimentation on animals, the research community aggressively argued against any change whatsoever in the use of animals in research, and thus for the continuation of an aggressively laissez-faire attitude toward animal use. On the other hand, radical activists argued for an immediate and abrupt cessation of animal use. The result, of course, was a stalemate, favoring the status quo.

4. In the same vein, both the new ethic being offered and the suggestions for reform entailed by it must accord with common sense and must be articulable in simple, ordinary language.

As mentioned earlier, it is not the case that historically philosophy did not *at all* engage the question of the moral status of animals. In a tradition most frequently associated with St. Thomas Aquinas and Immanuel Kant, and incorporated into the legal systems of most civilized societies beginning in the late eighteenth century, *cruelty* to animals was vigorously proscribed, though animals in themselves were denied moral status. The condemnation of such cruelty resulted from the realization that if people were permitted to be cruel to animals, as a matter of psychological fact, those who were would "graduate" to being cruel and abusive to people. A far more profound and intellectually bold move was that of utilitarian thinkers Jeremy Bentham and John Stuart Mill, who famously based candidacy for moral status on the ability to feel pleasure and pain. This approach was appropriated by Peter Singer in his revolutionary 1975 book *Animal Liberation,* the first contemporary attempt to ground full moral status for animals.

As articulated by Aquinas and Kant, animals in themselves do not enjoy direct moral status. But allowing *cruelty* to animals has a pernicious psychological effect upon humans. Cruelty, in the legal system, is defined as infliction of "deviant, unnecessary, extraordinary, purposeless, intentional, sadistic pain and suffering on an animal, serving no legitimate purpose" and failure to "minister to the necessities of man," as one judge put it. Aquinas and Kant argued that if pathological people are allowed to be cruel to animals, they will eventually be led to abusing people, which is socially undesirable. This insight has been buttressed by a good deal of twentieth-century psychological and sociological research. Most serial killers have early histories of animal abuse. The vast majority of violent offenders in Leavenworth federal prison have early histories of animal abuse, as do students who open fire on their classmates. Animal abuse, along with bedwetting and starting fires, is considered one of the key signs of nascent psychopathy.

Why can we not broaden the anti-cruelty ethic to cover other animal treatment? It is because only a tiny percentage of animal suffering is the result of deliberate, sadistic cruelty. Cruelty, as descriptive of psychological deviance, would not cover animal suffering that results from nonpathological pursuits such as industrial agriculture, safety testing of toxic substances on animals, and all forms of animal research. People who raise animals for food in an industrial setting, or who do biomedical research on animals, or who run zoos are not driven by sadistic desires to hurt these creatures. Rather, they generally believe they are doing social good, providing cheap and plentiful food, or medical advances, or educational opportunities, and they are in fact traditionally so perceived socially. Of all the suffering that animals endure at human hands, only a tiny fraction, less than 1 percent, is the result of deliberate cruelty.

An additional flaw in the anti-cruelty ethic/laws is that they cover only physical harm. Psychological harm and torture can be far more devastating to an animal than a beating or other form of physical abuse, but such abuse is invisible to the cruelty laws, again pointing up their conceptual inadequacy. (A friend of mine rescued a young female bullmastiff from abusive yahoos, who would shoot her with paintballs each morning before feeding her. Even though the shots in themselves were not severely painful, the dog grew up in a state of constant fear, shying away from strangers, and remained this way for life.) These weaknesses notwithstanding, more than forty states have now raised animal cruelty from a misdemeanor to a felony, evidencing growing social concern for animal treatment.

This leaves utilitarianism as the source of the only clearly articulated basis for a robust animal ethic in the history of philosophy before the twentieth century. As noted above, Singer drew on the classical utilitarian writings of Bentham and Mill when he pioneered in publishing the first comprehensive book on animal ethics, *Animal Liberation,* in 1975. Bentham famously affirmed that

> other animals, which, on account of their interests having been neglected by the insensibility of the ancient jurists, stand degraded into the class of things. . . . The day has been, I grieve it to say in many places it is not yet past, in which the greater part of the species, under the denomination of slaves, have been treated . . . upon the same footing as . . . animals are still. The day may come, when the rest of the animal creation may acquire those rights which never could have been withholden from them but by the hand of tyranny. The French have already discovered that the blackness of skin is no reason why a human being should be abandoned without redress to the caprice of a tormentor. It may come one day to be recognized that the number of legs, the villosity of the skin, or the termination of the os sacrum, are reasons equally insufficient for abandoning a sensitive being to the same fate. What else is it that should trace the insuperable line? Is it the faculty of reason,

or perhaps, the faculty for discourse? . . . The question is not, Can they reason? Nor, Can they talk? but, Can they suffer? Why should the law refuse its protection to any sensitive being? . . . The time will come when humanity will extend its mantle over everything which breathes . . . (1996, chap. 17)

Since animals are capable of feeling pain and pleasure, they thus according to Bentham belong within the scope of moral concern. But I was not satisfied with a utilitarian basis for animal ethics. First of all, anyone not accepting a utilitarian underpinning for ethics in general would not accept the resulting animal ethics. Second, basing ethics on maximizing pleasure and minimizing pain across a society presupposes the commensurability of all forms of pleasure and suffering. I could never understand how such disparate forms of negative experience as isolation from mother, hot-iron branding, neglect, beating, lack of affection, being yelled at, being burned, being deprived of food or water, and being denied interaction with conspecifics or the full range of positive experiences could be neatly laid out on a homogeneous scale, measured, and compared. Further, it seemed to follow from utilitarian principles that if some act produces more pleasure than pain, however heinous the act, it was morally acceptable. If we could alleviate the pain of 2,000 burn victims by inflicting burn pain on 1,900 experimental subjects, human or animal, such experimentation becomes ipso facto morally acceptable, as does torturing a person who has placed a bomb that must be defused somewhere in an elementary school. The entire wrongness of minority oppression for the benefit of the majority, and the question of the rights accorded to these minorities, simply vanishes under the pressure of focusing on "the greatest good for the greatest number," which can in turn lead to totalitarian oppression.

There are many other philosophical objections to accepting utilitarianism as a basis for all ethics, including animal ethics. But that is not the chief obstacle facing grounding animal ethics on a utilitarian foundation. The chief problem lies in the fact that utilitarianism is simply not universally accepted even by everyone seeking a theoretical basis for ethics, let alone by most ordinary citizens, who need to accept animal ethics to make it practicable. Here is the critical point: If the given basis for animal ethics does not compel the allegiance of the vast majority of people it addresses, it becomes more like religion, open to indefinite diversity, rather than being adopted by an overwhelming consensus. And it appears that no theoretical basis for animal ethics to be found in the history of thought has anything like the persuasive power to compel such adherence.

Let me illustrate this point anecdotally. At one point in my career, I was engaged in conversation with a colleague of mine from Korea who taught Eastern thought and Asian religion. As he often did, he was complaining about our faculty salaries and reflecting on possible sources of additional income. Only half joking, I made the following suggestion based on the fact that he was quite charismatic and did a good job playing the "Asian sage" role. "Why don't you start a cult or religion? If you worked diligently, you should be able to attract a couple of thousand adherents. If people can buy into Raelianism, Scientology, and the Maharishi, you can surely sell your version of Eastern thought. Part of the dogma you teach could be the renunciation of personal wealth and property, to be turned over to you by potential acolytes. It would be easy to find a thousand people willing to believe in virtually anything. If you charge each one a measly $10,000 to undertake a spiritual journey under your guidance, you will quickly amass $10 million and no longer need to rely on your university salary."

I was, of course, being facetious. Yet it is precisely in this spirit that too many philosophers approach the extraordinarily difficult task of deriving an ethic for guiding and constraining the treatment of animals in society. Like pure mathematicians, they build perfectly consistent systems that are internally logically sound and even aesthetically appealing but have no contact with reality. It is eminently reasonable for the mathematician to proceed in this way; pure mathematics has a life of its own with no need to be interpretable in a way that fits the real world. In ethics, however, a moral system that does not mesh with reality is of little value in enhancing the treatment of those it purports to cover. An impracticable ethic, unlike an impossible mathematical universe, is not beautiful; it is more silly than anything else.

For example, Singer himself has argued, for utilitarian reasons, that the only way to ameliorate the suffering of farm animals raised in industrial animal factories is to stop eating meat and adopt a vegetarian if not vegan diet. A moment's reflection reveals the implausibility of that suggestion. People will not give up steaks, hot dogs, and hamburgers even when counseled to do so by their physicians to improve their own health or even to save their own lives, so the chances that they will do so in the face of a philosophical argument are vanishingly small. In other words, not only must a successful animal ethic be logically consistent and persuasive, it must also be seen as practicable and it must suggest real solutions that people can both advocate and adhere to. As we will see later in our discussion, the elimination of tiny cages for sows represents a major improvement in pig welfare, yet was not difficult to effect, despite requiring major changes in the swine industry.

Social, Personal, and Professional Ethics

PETER SINGER'S WORK was a major step forward, but the problems I just enumerated remained. It appeared that creating a new ethic for animals was at an impasse, when suddenly I had a revelation. In the course of thinking through the veterinary ethics course I was committed to teaching, I realized that there is an ethic as well as an ethical theory to which the vast majority of citizens subscribe. It is the ethic according to which we in society are held accountable. It is what I call the social-consensus ethic.

There are two very different senses of "ethics" that are often confused and conflated and that must be distinguished at the outset to allow for viable discussion of these matters. The first sense of ethics I shall call ethics$_1$. In this sense ethics is the set of principles or beliefs that govern views of right and wrong, good and bad, fair and unfair, just and unjust. Whenever one asserts that "killing is wrong," or that "discrimination is unfair," or that "one oughtn't belittle a colleague," or that "it is laudable to give to charity," or that "abortion is murder," one is explicitly or implicitly appealing to ethics$_1$—moral rules that one believes ought to bind society, oneself, and/or some subgroup of society, such as veterinarians.

Under ethics$_1$ must fall a distinction between social ethics, personal ethics, and professional ethics. Of these, social ethics is the most basic and most objective, in a sense to be explained shortly. People, especially scientists, are tempted sometimes to assert that unlike scientific judgments, which are "objective," ethical judgments are "subjective" opinion and not "fact," and thus they are not subject to rational discussion and adjudication. Although it is true that one cannot conduct experiments or gather data to decide what is right and wrong, ethics, nevertheless, cannot be based upon personal whim and caprice. If anyone doubts this, let that person go out and rob a bank in front of witnesses, then argue before a court that, in his or her ethical opinion, bank robbery is morally acceptable if one needs money.

In other words, the fact that ethical judgments are not validated by gathering data or doing experiments does not mean that they are simply a matter of individual subjective opinion. If one stops to think about it, one will quickly realize that in real life very little socially important ethics is left to one's subjective opinion. Consensus rules about rightness and wrongness of actions that have an impact on others are in fact articulated in clear social principles, which are in turn encoded in laws and policies. All public regulations, from the zoning of pornographic bookstores outside of school zones to laws against insider trading and murder, are examples of consensus ethical principles "writ large," in Plato's felicitous phrase, in public policy. This is not to say that, in every case, law and ethics are congruent—we can all think of examples of things that are legal yet generally considered immoral (tax dodges for the ultra-wealthy, for instance) and of things we consider perfectly moral that are illegal (parking one's car for longer than two hours in a two-hour zone).

But, by and large, if we stop to think about it, there must be a pretty close fit between our morality and our social policy. When people attempt to legislate policy that most people do not consider morally acceptable, the law simply does not work. A classic example is, of course, Prohibition, which did not stop people from drinking, but rather funneled the drinking money away from legitimate business to bootleggers. So there must be a goodly number of ethical judgments in society that are held to be universally binding and socially objective. Even though such judgments are not objective in the way that the belief that "water boils at 212 degrees Fahrenheit" is objective (that is, they are not validated by the way the world works), they are nonetheless objective as rules governing social behavior. We are all familiar with other instances of this kind of objectivity. For example, it is an objective rule of English that one cannot say, "You ain't gonna be there." Though people, of course, do say it, it is *objectively* wrong to do so. Similarly, the bishop in chess can objectively move only on diagonals of its own color. Someone may, of course, move the bishop a different way, but that move is objectively wrong, and that person would then not be "playing chess."

Those portions of ethical rules that we believe to be universally binding on all members of society, and socially objective, I will consider part of the *social-consensus* ethic. A moment's reflection reveals that without some such consensus ethic, we could not live together: we would have chaos and anarchy, and society would be impossible. This is true for any society at all that intends to persist—there must be rules governing everyone's behavior, and they must be objectively encoded in laws. Do the rules need to be the same for all societies? Obviously not—we all know that there are endless ethical variations across societies. Does there need to be at least a common core in all these ethics? That is a rather profound question I shall address later. For the moment, however, we all need to agree that there exists an identifiable social-consensus ethic in our society by which we are all bound.

Now, the social-consensus ethic does not regulate all areas of life that have ethical relevance—certain areas of behavior are left to the discretion of the individual, or, more accurately, to his or her *personal* ethic. Such matters as what one reads, what religion one practices or does not practice, and how much charity one gives and to whom are all matters left in our society to one's personal beliefs about right and wrong and good and bad. This has not always been the case, of course; all these examples, during the Middle Ages, were appropriated by a theologically based social-consensus ethic. And this fact illustrates a very important point about the relationship between social-consensus ethics and personal ethics: as a society evolves and changes over time, certain areas of conduct may move from the concern of the social-consensus ethic to the concern of the personal ethic, and vice versa.

An excellent example of a matter that has recently moved from the concern of the social ethic, and from the laws that mirror that ethic, to the purview of the personal ethic is the area of sexual behavior. Whereas once laws constrained activities like homosexual behavior, adultery, and cohabitation, these things are now increasingly left to one's personal ethic in Western democracies. With the advent during the 1960s of the view that sexual behavior that does not hurt others is not a matter for social regulation but, rather, for personal choice, social regulation of such activity withered away. Some years ago the mass media reported, with much hilarity, that there was still a law on the books in Greeley, Colorado, a university town, making cohabitation a crime. Radio and TV reporters chortled as they remarked that, if the law were to be enforced, a goodly portion of the Greeley citizenry would have to be jailed. And all of us living today realize that homosexual preferences are rapidly moving away from social or legal condemnation—witness the increasing social and legal acceptance of gay marriage.

On the other hand, we must note that many areas of behavior once left to one's personal ethic have since been appropriated by the social ethic. When I was growing up, paradigm cases of what society left to one's personal choice were represented by the kind of person to whom one chose to rent or sell one's real property and whom one hired for jobs. The prevailing attitude was that these decisions were your own damn business. This, of course, is no longer the case. Federal law now governs renting and selling of property and hiring and firing.

Generally, as such examples illustrate, conduct becomes appropriated by the social-consensus ethic when how it is dealt with by personal ethics is widely perceived to be *unfair or unjust.* The widespread failure to rent to, sell to, or hire minorities, which resulted from leaving these matters to individual ethics, evolved into a situation viewed by society as unjust, and this led to the passage of strong social-ethical rules against such unfairness. As we shall see, the treatment of animals in society is also moving into the purview of the social-consensus ethic, as society begins to question the injustice that results from leaving such matters to individual discretion.

The third component of ethics$_1$, in addition to social-consensus ethics and personal ethics, is *professional* ethics. Members of a profession are first and foremost members of society—citizens—and thus are bound by all aspects of the social-consensus ethic not to steal, murder, break contracts, and so on. However, professionals—be they physicians, attorneys, or veterinarians—also perform specialized and vital functions in society. This kind of role requires special expertise and special training, and involves special situations that ordinary people do not face. The professional functions that physicians and veterinarians perform also warrant special privileges—for example, dispensing medications and performing surgery. Democratic societies have been prepared to give professionals some leeway and to assume that, given the technical nature of professions and the specialized knowledge their practitioners possess, professionals will understand the ethical issues they confront better than society does as a whole. Thus society generally leaves it to such professionals to set up their own rules of conduct. In other words, the social ethic offers general rules, creating the stage on which professional life is played out, and the subclasses of society comprising professionals are asked to develop their own ethic to cover the special situations they deal with daily. In essence, society says to professionals, "Regulate yourselves the way we would regulate you if we understood enough about what you do to regulate you." Because of this situation, professional ethics occupies a position midway between social-consensus ethics and personal ethics, because it neither applies to all members of society nor are its main components left strictly to individuals. It is, for example, a general rule of human medical ethics for psychiatrists not to have sex with their patients.

The failure of a profession to operate in accordance with professional ethics that reflect and are in harmony with the social-consensus ethic can result in a significant loss of autonomy by the profession in question. One can argue, for example, that recent attempts to govern health care by legislation are a result of the human medical community's failure to operate in full accord with the social-consensus ethic. When hospitals turn away poor people or aged stroke victims, or when insurance companies fail to provide coverage for preexisting conditions, or when pediatric surgeons fail to use anesthesia on infants or give less analgesia to adolescents than to adults with the same lesion (Rollin, 2007), they are not in accord with social ethics, and it is only a matter of time before society will appropriate regulation of such behavior. In veterinary medicine, growing social awareness of the irresponsible use and dispensing of pharmaceuticals more than forty years ago (use of antimicrobials in farm animals for growth promotion, which in effect bred for pathogen resistance to these drugs, creating a threat to humans) threatened the privilege of veterinarians to prescribe drugs in an extra-label fashion (i.e., in a way not dictated by the manufacturer)—a privilege whose suspension would have in a real sense hamstrung veterinarians. Because so few drugs are approved for animals, veterinary medicine relies heavily on extra-label drug use. The issue of antibiotic overuse in agriculture is still with us—the majority of antibiotics manufactured in the United States are used in food animals to enhance growth or to stifle the negative effects of the crowded and pathogenic conditions under which farm animals are raised. When scientists at Johns Hopkins University were studying water samples in the Delmarva region of the United States (Delaware, Maryland, Virginia), they were shocked to find Vancomycin, a cutting-edge human antibiotic, in those samples, as the drug that should have been saved only for human use in cases of pathogens resistant to "standard" antibiotics was being used to make factory farming of poultry possible.

Thus far we have looked at ethics$_1$—the set of principles that govern people's views of right and wrong, good and bad, fair and unfair, just and unjust—and found that it can be further divided into social-consensus ethics, personal ethics, and professional ethics. Now we must consider a less familiar, secondary notion. Ethics$_2$ is the logical, rational study and examination of ethics$_1$, which may include attempting to justify the principles of ethics$_1$, seeking out inconsistencies in the principles of ethics$_1$, drawing out ethics$_1$ principles that have been hitherto ignored or unnoticed, engaging the question of whether all societies ought ultimately to have the same ethics$_1$, and so on. This secondary sense of ethics—ethics$_2$—is thus a branch of philosophy. Most of what we are doing in this book is ethics$_2$, examining the logic of ethics$_1$. Socrates's activities in ancient Athens were a form of ethics$_2$. Whereas we in society learn ethics$_1$ from parents, teachers, churches, movies, books, peers, magazines, newspapers, and mass media, we rarely learn to engage in ethics$_2$ in a disciplined, systematic way unless we take an ethics class in a philosophy curriculum. In one sense this is fine—vast numbers of people are diligent practitioners of ethics$_1$ without ever engaging in ethics$_2$. On the other hand, failure to engage in ethics$_2$—rational criticism of ethics$_1$—can lead to incoherence and inconsistencies in ethics$_1$ going unnoticed, unrecognized, and uncorrected. Although not everyone needs to engage in ethics$_2$ on a regular basis, there is value in at least some people monitoring the logic of ethics$_1$, be it social-consensus ethics, personal ethics, or professional ethics. Such monitoring helps us detect problems that have been ignored or have gone undetected and helps us make ethical progress.

As I have indicated briefly, ethical theories determine what falls within the arena of moral deliberation. But ethical theories also serve to decide between conflicting ethical principles. So let us now look at how individuals can rationally make ethical decisions and how they can rationally convince (or attempt to convince) others with whom they have putative disagreement. In the first place, one must attempt to define all ethically relevant components of the situation. Assuming that the situation is thus analyzed and the answer is not dictated by the social ethic, what does one do next? Let us here take a hint from the philosopher Ludwig Wittgenstein and ask ourselves how we learned about right and wrong, good and bad. As children, we might, for example, reach over to steal our brother's chocolate pudding and be told by our mother, "No! That is wrong!" In other words, this is how we learn that certain actions are wrong, or that they are right, as when we are praised for sharing our chocolate pudding with our brother. As we get older, we gradually move from learning that forcibly taking the chocolate pudding away from brother is wrong on this particular occasion, to the generalization that taking it on any occasion is wrong, to the more abstract generalization that taking something from someone else without permission is wrong, to the even more abstract notion that stealing is wrong. In other words, we ascend from particulars to generalizations in our moral beliefs, just as we do in our knowledge of the world, moving from "Don't touch this radiator," to "Don't touch any hot objects," to "Hot objects cause burns if touched."

Let us call the ethical generalizations that we learn as we grow moral principles (or ethics$_1$ principles). Although we originally learn such moral principles primarily from our parents, as we grow older we acquire them from many and varied sources—friends and other peers, teachers, churches, movies, books, radio and television, newspapers, magazines, and so on. We learn such diverse principles as "It is wrong to lie," "It is wrong to steal," "It is wrong to hurt people's feelings," "It is wrong to use drugs," "Stand up for yourself," and, of course, many others. Eventually, we have the mental equivalent of a hall closet chock-full of moral principles, which we (ideally) pull out in the appropriate circumstances. So far this sounds simple enough. The trouble is that sometimes two or more principles fit a situation yet patently contradict one another. It is easy to envision a multitude of situations wherein this dilemma might occur.

For example, we have all learned the principles not to lie and not to hurt others' feelings. Yet these may stand in conflict in social-ethical situations, as when a coworker or my wife asks me, "What do you think of my new three-hundred-dollar hairdo?" and I think it is an aesthetic travesty. Similarly, many of my male students who grew up on ranches face such tension. On the one hand, they have been brought up as Christians and taught the principle "Turn the other cheek." On the other hand, they have also been taught "not to take any crap and to stand up for yourself." As a third example, a female colleague tells of suffering a great deal of anguish when dating, as she had been taught both to be chaste and not to make others feel bad. Principles also contradict each other in some professional-ethical situations. Veterinarians, like all professionals, face conflicting principles— indeed, one need go no further than the traditional veterinarian's oath to see a clear conflict. For example, there is certainly a tension between the injunction to "advance scientific knowledge" and the injunction to "ameliorate animal suffering," as scientific knowledge often advances by creating animal suffering.

When faced with such conflicts, many of us simply do not notice them. As one of my cowboy students said to me once about the internal conflict between turning the other cheek and not being bullied: "What's the problem, Doc? 'Turn the other cheek' comes out in church, the other one comes out in bars." Obviously, this response is less than satisfactory!

The key to resolving such contradictions lies in how one prioritizes the principles in conflict. Obviously, if they are given equal priority, one is at an impasse. So we need a higher-order theory to decide which principles are given greater weight in which sorts of situations and to keep us consistent in our evaluations so that we do the same sort of prioritizing in situations that are analogous in a morally relevant way. In this regard one can perhaps draw a reasonable analogy between levels of understanding in science (that is, knowledge of the world) and ethics. In science one begins with individual experiences (for example, of a moving body), one then learns a variety of laws of motion (celestial motion, Kepler's laws, terrestrial motion), and one finally unifies the variegated laws under one more general theory from which they can all be derived (Newton's theory of universal gravitation). Similarly, in ethics one begins with awareness that particular things are wrong (or right), moves to principles, and then ascends to a theory that prioritizes, explains, or provides a rationale for both having and applying the principles. Theories can also help us identify ethical components of situations wherein we intuitively surmise there are problems but cannot sort them out.

Construction of such ethical theories has occupied philosophers from Plato to the present. It is beyond the scope of this discussion to survey the many diverse theories that have been promulgated. But it is valuable to look at the two significantly different systems I have already alluded to that nicely represent extremes in ethical theory and that, more important, have been synthesized in the theory underlying our own social-consensus ethic.

Ethical theories tend to fall into two major groups—those stressing goodness and badness, that is, the results of actions, and those stressing rightness and wrongness, or duty, that is, the intrinsic properties of actions. The former are called consequentialist, or teleological, theories (from the Greek word *telos* (pl. *teloi*), meaning "result," "end," or "purpose"). The latter are termed deontological theories (from the Greek word *deontos*, meaning "necessity" or "obligation")—in other words, what one is obliged to do. The most common deontological theories are theologically based, wherein action is obligatory because commanded by God.

What is our social-consensus ethic, and what is the theory underlying it? It is basically the ethic encoded and articulated in the US Constitution and the laws historically derivative therefrom and, with some variation, in laws in other Western democratic societies. To understand this ethic, we need to contrast the major historical opposing ethical theories we cited briefly earlier in our discussion.

The most well-known consequentialist theory is utilitarianism. It has appeared in a variety of forms throughout history but is most famously associated with nineteenth-century philosophers Jeremy Bentham and John Stuart Mill (Bentham, 1996; Mill, 1902), as discussed in the previous chapter. In its simplest version, utilitarianism holds that one acts in given situations according to what produces the greatest happiness for the greatest number, wherein happiness is defined in terms of pleasure and absence of pain. The principles of utilitarianism are generalizations about courses of action that tend to produce more happiness than unhappiness. In situations wherein principles conflict, one decides which course of action to take by calculating which is likeliest to produce the greatest happiness. Thus, in the trivial case of someone asking me what I think of her new hairstyle, which I find repulsive but where I do not wish to hurt her feelings, telling a "little white lie" will likely produce no harm, whereas telling the truth will result in hostility and bad feeling, so one ought to choose the former course of action. As we saw, Bentham and Singer based animal ethics on utilitarian theory.

There are many problems with this sort of theory, but they lie beyond the scope of this discussion. The only point relevant here is that adherence to such a theory resolves conflict among principles like "Do not lie" and "Do not hurt anyone's feelings" by providing a higher-order rule for decision making.

Those of us who grew up with very liberal parents will quickly recognize the utilitarian approach. Suppose you approach such parents in a quandary. You are thinking of entering into an adulterous relationship with a married woman. You explain that she is terminally ill and, despised and abandoned by her vile, abusive husband who does not care what she does but who nonetheless is sadistically blocking a divorce, she is attempting to snatch a brief period of happiness before her demise. These parents might well say, "Adultery is generally wrong, as it usually results in great unhappiness. But in this case perhaps you both deserve the joy you can have together. . . . No one will be hurt."

On the other hand, those who grew up with German Lutheran parents can imagine a very different scenario if one approached them with the same story. They would be very likely to say, "I don't care what the results will be—adultery is always wrong! Period!" This is, of course, a strongly deontological position. The most famous rational reconstruction of such a position is to be found historically in the German philosopher Immanuel Kant's *Foundations of the Metaphysics of Morals*. According to Kant, ethics is unique to rational beings. Rational beings, unlike other beings, are capable of formulating universal truths of mathematics, science, and so on. Animals, lacking language, simply do not have the mechanism to think in terms such as "all X is Y." As rational beings, humans are bound to strive for rationality in all areas of life. Rationality in the area of conduct is to be found in subjecting the principle of action you are considering to the test of universality, by thinking through what the world would be like if everyone behaved the way you are considering behaving. Kant called this requirement the Categorical Imperative, that is, the requirement of all rational beings to judge their intended actions by the test of universality. In other words, suppose you are trying to decide whether you should tell a little white lie in an apparently innocuous case, like the ugly hairdo dilemma. Before doing so, you must test that action by the Categorical Imperative, which enjoins you to "act in such a way that your action could be conceived to be a universal law." So before you lie, you conceive of what would occur if everyone were allowed to lie whenever it was convenient to do so. In such a world the notion of telling the truth would cease to have meaning, and thus so, too, would the notion of telling a lie. In other words, no one would trust anyone.

Thus, universalizing a lie leads to a situation that destroys the possibility of the very act you are contemplating, and therefore becomes rationally indefensible, *regardless of the good or bad consequences in the given case*. By the same token, subjecting your act of adultery to the same test shows that if one universalizes adultery, one destroys the institution of marriage, thereby in turn rendering adultery impossible. Thus, in a situation of conflicting principles, one rejects the choice that could not possibly be universalized.

Kant goes on to draw some other implications from his account, including the conclusion that one should always treat other rational beings as "ends in themselves, not merely as means," but these are irrelevant to our example. While Kant's theory also is open to some strong criticisms, these too need not be discussed here. The point is that both personal and social ethics must be based in some theory that prioritizes principles to assure consistency in behavior and action. Having such a theory helps prevent arbitrary and capricious actions.

Whatever theory we adhere to as individuals, we must be careful to assure that it fits the requirements demanded of morality in general: it must treat people who are relevantly equal equally; it must treat relevantly similar cases the same way; it must avoid favoring some individuals for morally irrelevant reasons (such as hair color); it must be fair, and not subject to whimsical change.

Obviously, a society needs some higher-order theory underlying its social-consensus ethic. Indeed, such a need is immediately obvious as soon as one realizes that every society faces a fundamental conflict of moral concerns—the good of the group or state or society versus the good of the individual. This conflict is obvious in almost all social decision making, be it the military's demanding life-threatening service from citizens or the legislature's redistributing wealth through taxation. It is in society's interest to send you to war—it may not be in yours, as you risk being killed or maimed. It is in society's interest to take money from the wealthy to support social programs or, more simply, to improve quality of life for the impoverished, but it arguably does not do the wealthy individual much good.

Different societies have of course constructed different theories to resolve this conflict. Totalitarian societies have taken the position that the group, or state, or Reich, or however they formulate the corporate entity, must unequivocally and always take precedence over the individual. The behavior of the Soviet Union under Stalin, Germany under Hitler, China under Mao, and Japan under the emperors bespeaks the primacy of the social body over individuals. On the other end of the spectrum are anarchistic communes, such as those of the 1960s, that give total primacy to individual wills and see the social body as nothing more than an amalgam of individuals. Obviously, societies along the spectrum are driven by different higher-order theories.

How does the ethic operative in our and other democratic societies resolve the tension between the individual and society? In my view, the United States has developed the best mechanism in human history for maximizing both the interests of the social body and the interests of the individual. Although we make most of our social decisions by considering what will produce the greatest benefit for the greatest number, a utilitarian/teleological/consequentialist ethical approach, we skillfully avoid the "tyranny of the majority," or the submersion of the individual under the weight of the general good. We do this by considering the individual as, in some sense, inviolable. Specifically, we consider those traits of an individual that we believe are constitutive of his or her *human nature*—what Aristotle called "telos"—to be worth protecting at almost all costs. We believe that individual humans are by nature thinking, speaking, social beings who do not wish to be tortured, want to believe as they see fit, desire to speak their minds freely, have a need to congregate with others of their choice, seek to retain their property, and so forth. We take the human interests flowing from this view of human nature as embodied in individuals and build protective legal/moral fences around them that insulate those interests even from the powerful, coercive effect of the general welfare. These protective fences guarding individual fundamental human interests even against the social interest are called rights. Not only do we as a society respect individual rights, we do our best to sanction other societies that ride roughshod over them.

In essence, then, the theory behind our social ethic represents a middle ground or synthesis between utilitarian and deontological theories. On the one hand, social decisions are made and conflicts resolved by appeal to the greatest good for the greatest number. But in cases wherein maximizing the general welfare could oppress the basic interests constituting the humanness of individuals, general welfare is checked by a deontological theoretical component, namely, respect for the individual human's nature—telos—and the interests flowing therefrom, which are in turn guaranteed by rights.

Before demonstrating how one can utilize this notion of a social-consensus ethic to introduce animals into the moral arena, one conceptual obstacle needs to be overcome. This is the ancient objection from ethical relativism, which asserts that all ethical positions are equally valid and that no society (or individual) can be said to have a better ethic than another. The Sophists in ancient Greece, for example, would point out that although incest was a heinous moral offense in Greece, it was the rule among the Egyptian royal family. One still finds this relativistic position among college freshmen and among scientists who see ethics as "opinion" rather than fact.

There are many refutations of relativism; here we will consider two.

1. Relativism is self-defeating: Relativism asserts that all ethical positions are equally valid or true. In saying this, the relativist admits that his own position has no special validity and that the ethical position that *denies* the legitimacy of relativism is as true as relativism. Thus, if relativism is correct, its absolute correctness cannot be asserted by its defenders.

2. There exist criteria for judging competing ethics: It is certainly true that there are differences in people's (and societies') ethical approaches. That in itself, however, does not mean that all approaches are equally valid. Perhaps we can judge different ethical views by comparing them to the basic purposes of ethics and to the reasons that there is a need for ethics in the first place.

As mentioned early in this discussion, rules for conduct are necessary if people are to live together—which of course they must. Without such rules, with people doing whatever they wish, chaos, anarchy, and what Thomas Hobbes called "the war of each against all" would ensue. Some idea of what such a world would be like may be gleaned from what happens during wars, floods, blackouts, and other natural or man-made disasters. A perennial source of fiction and drama, such situations lead to looting, pillaging, rape, robbery, outrageous black-market prices for such necessities as food, water, and medicine, and so on.

What sorts of rules best meet the needs dictated by social life? We know through ordinary experience and common sense what sorts of things matter to people. Security regarding life and property is one such need. The ability to trust what others tell us is another. Leaving certain things in one's life to one's own choices is a third. Clearly, certain moral constraints, principles, and even theories will flow from these needs. Rational self-interest dictates that if I do not respect your property, you will not feel any need to respect mine. Because I value my property and you value yours, and we cannot stand watch over our property all the time, we "agree" not to steal and adopt this agreement as a moral principle. A similar argument could be mounted for prohibitions against killing, assault, and so on. By the same token, the prohibition against lying could be based naturally in the fact that communication is essential to human life and that a presupposition of communication is that, in general, the people with whom one is conversing are telling the truth.

By the same token, as we have seen Kant emphasize, certain conclusions can be drawn about morality from the fact that it is based on reason. We would all agree that the strongest way someone can err rationally is to be self-contradictory. To be sensible, or rational, we must be consistent. According to some thinkers, something very like the Golden Rule is a natural consequence of a requirement for consistency. In other words, I can be harmed in certain ways, helped in others, and wish to avoid harm and fulfill my needs and goals—and I see precisely the same features in you and the same concerns. Thus, if I believe something should not be done to me, I am led by the similarities between us to conclude that neither should it be done to you, by me or by any other human being. In fact, it is precisely to circumvent this plausible sort of reasoning that we focus on differences between ourselves and others: color, place of origin, social station, heritage, genealogy, anything that might serve to differentiate you from me, us from them, so I do not have to apply the same concerns to others as to me and mine. The history of civilization, in a way, is a history of discarding differences that are not relevant to how one should be treated, like sex or skin color. In sum, some notion of justice—equals should be treated equally—has been said to be a simple deduction from logic.

In support of this argument, one can say that at least a core of common principles survives even cross-cultural comparison. For example, some version of the Golden Rule can be found in Judaism, Christianity, Islam, Brahmanism, Hinduism, Jainism, Sikhism, Buddhism, Confucianism, Taoism, Shintoism, and Zoroastrianism. And it stands to reason that certain moral principles would evolve in all societies as a minimal requirement for living together. Any society with property would need prohibitions against stealing; communication necessitates prohibitions against lying; murder could certainly not be freely condoned; and so on. In any case, even if one is philosophically drawn to relativism, one must, like it or not, obey the social-consensus ethic.

There is a far more audacious argument against relativism and for favoring the American democratic ethic as the best social-consensus ethic. What makes this so audacious is that we live in an era where political correctness is dominant and tends to eclipse any position not wholly committed to diversity and multiculturalism. Although I risk transgressing current political correctness ideology, I would affirm that it does make perfect sense to say that certain cultures' moral systems are better than others'. This seems intuitively obvious. Very few of us are prepared to say that cultures that perform clitoral mutilations on female children, or practice infanticide, or seize property arbitrarily, or put people in concentration camps, or permit rape, or discriminate on the basis of skin color are morally as good as cultures that eschew such practices. Intuitions, however, are not arguments. But I believe we can, using the insights of philosophers like Plato, Hobbes, and John Rawls, develop an argument justifying our intuitions.

Plato, at the end of *The Republic,* affirms that only a person who has grasped genuine, absolute morality can make a wise and rational decision about what sort of life to choose if one is reincarnated. One way of interpreting this Platonic myth or allegory is to say that only when one understands the role of ethical systems can one be in a position to know what systems are better than others. As we discussed earlier, one manifest purpose of ethical systems is to facilitate people's living together effectively, since humans are social from birth. We also saw that, as Hobbes says, a primordial function of ethics is protecting people from each other, and another function, in a more positive vein, is facilitating cooperative efforts. If people willingly embrace a system of ethical rules, it must be in part because they are better off doing so than not doing so, and because they see those rules as a reasonable mechanism for *fairly* distributing the benefits and costs implicit in social life. Few of us would think an ethical system fair if it heavily favored one skin color, religion, or ancestry rather than accomplishment, or talent and achievement.

Let us return to Plato's case: You are asked to choose in what ethical system you wish to be reincarnated. Additionally, as Rawls has beautifully argued in his *Theory of Justice,* let us assume that you do not know ahead of time your role in the system you choose. Whereas many of us would choose to live in prerevolutionary France as an aristocrat, or in ancient Greece as a citizen, few would choose those systems if we knew we would be slaves. If faced with such a choice, it would surely be most rational to choose a society where your fundamental interests were protected as much as possible regardless of your station in society, regardless of whether you were rich or poor, noble or commoner, white or black, male or female. Further, any rational person would tend to choose a society wherein one's fate is determined by one's achievements and abilities, not by the family or class or caste one is born into. If we look historically at the vast panoply of moral systems serving as the consensus ethics of different societies, we would have to conclude that current democratic societies represent the least arbitrary and fairest moral systems, wherein one's fate is mostly in one's own hands rather than being determined by irrelevant features such as accidents of birth and which seem to be continuing to evolve in the direction of greater fairness. For this reason, I am quite comfortable emphatically rejecting ethical relativism. Even though our society and its ethics have, at various points in our history, been guilty of major arbitrary discrimination and unfairness, the crucial point is that our societal ethic contains within it the seeds of and mechanisms for correction and transformation, allowing us to overcome that injustice.

It is very likely that there has been more and deeper social-ethical change since the middle of the twentieth century than occurred during centuries of an ethically monolithic period such as the Middle Ages. Anyone over forty has lived through a variety of major moral earthquakes aimed at eliminating injustice and unfairness: the sexual revolution, the civil rights movement and the end of socially sanctioned racism, the banishing of IQ differentiation, the rise of homosexual militancy, the end of administrators' acting *in loco parentis* in universities, the advent of consumer advocacy, the end of a mandatory retirement age in most fields, the mass acceptance of environmentalism, the growth of a "sue the bastards" mind-set, the implementation of affirmative action programs, the rise of massive drug use, the designation of alcoholism and child abuse as diseases rather than moral vices, the rise of militant feminism, the emergence of sexual harassment as a major social concern, the demands by the handicapped for equal access, the rise of public suspicion of science and technology, the mass questioning of animal use in science and industry, the end of colonialism, the rise of political correctness. All these are patent examples of the magnitude of ethical change during this brief period and of our genuine social commitment to advancing a more just society.

Reminding versus Teaching

THE VERY INTERESTING question that now arises is this: How does ethical change in individuals, subgroups of society, and society as a whole occur? As is well known, moral judgments are not verified or falsified by reference to experiment or to new data gathered about the world—indeed, recognition of this fact has led twentieth-century science to conclude erroneously that science is "value-free" in general and "ethics-free" in particular. In any event, the knowledge that ethics is not validated by gathering empirical information has led some people to conclude that the only way to change anyone's (or any society's) ethical beliefs is by emotion and propaganda—and that reason has no role.

The best account of the subtle way in which ethical change occurs in a rational manner is given by Plato in the dialogue *Meno*. Plato explicitly states that people who are attempting to deal with ethical matters rationally cannot *teach* rational adults, they can only *remind* them. Whereas one can teach one's veterinary students the various parasites of the dog and demand that they spit back the relevant answers on a quiz, one cannot do that with matters of ethics, except insofar as one is testing their knowledge of the social ethic as objectified in law—what they may not do with drugs, for example. (Children, of course, *are* taught ethics.)

Some years ago I experienced an amusing incident that underscores this point. That year I had a class of particularly obstreperous veterinary students. Throughout the course they complained incessantly that I was only raising ethical questions, not giving them "answers." One morning I came to class an hour early and filled the blackboard with a variety of maxims, such as, "Never euthanize a healthy animal"; "Always tell the whole truth to

31

clients"; "Don't castrate without anesthesia"; "Don't dock tails or crop ears"; and so on. When the students filed into class, I told them to copy down these maxims and memorize them. "What are they?" they asked. "These are the answers," I replied. "You've been badgering me all semester to give you answers; there they are." "Who the hell are you to give us answers?" they immediately chorused.

This illustrates the first part of Plato's point, that one cannot teach ethics to rational adults the same way one teaches state capitals. But what of his claim that though one cannot teach, one can remind?

In answering this question, I always appeal to a metaphor from the martial arts. One can, when talking about physical combat, distinguish between sumo and judo. Sumo involves two large men trying to push each other out of a circle. If a one-hundred-pound man is engaging a four-hundred-pound man in a sumo contest, the result is a foregone conclusion. In other words, if one is simply pitting force against force, the greater force will prevail. On the other hand, a one-hundred-pound man can fare quite well against a four-hundred-pound man if the former uses judo, that is, turns the opponent's force against him. For example, you can throw much larger opponents simply by "helping them along" in the direction of their attack on you.

When you are trying to change people's ethical views, you accomplish nothing by clashing your views against theirs—all you get is a counterthrust. Far better to show that the conclusion you wish them to draw is implicit in what *they* already believe, albeit unnoticed. This is the sense in which Plato talked about "reminding."

As one who spends a good deal of my time attempting to explicate the new ethic for animals to people whose initial impulse is to reject it, I can attest to the futility of ethical sumo and the efficacy of moral judo. One excellent example leaps to mind. Some years ago I was asked to speak at the Colorado State University Rodeo Club about the new ethic in relation to rodeo. When I entered the room, I found some two dozen cowboys seated as far back as possible, cowboy hats over their eyes, booted feet up, arms folded defiantly, arrogantly smirking at me. With the quick-wittedness for which I am known, I immediately sized up the situation as a hostile one.

"Why am I here?" I began by asking. No response. I repeated the question. "Seriously, why am I here? You ought to know, you invited me."

One brave soul ventured, "You're here to tell us what is wrong with rodeo."

"Would you listen?" said I.

"Hell no!" they chorused.

"Well, in that case I would be stupid to try, and I'm not stupid."

A long silence followed. Finally someone suggested, "Are you here to help us think about rodeo?"

"Is that what you want?" I asked.

"Yes," they said.

"Okay," I replied, "I can do that."

For the next hour, without mentioning rodeo, I discussed many aspects of ethics: the nature of social morality and individual morality, the relationship between law and ethics, the need for an ethic for how we treat animals. I queried the cowboys as to their position on the latter question. After some dialogue they all agreed that, as a minimal ethical principle, one should not hurt animals for trivial reasons. "Okay," I said. "In the face of our discussion, take a fifteen-minute break, go out in the hall, talk among yourselves, and come back and tell me what *you guys* think is wrong with rodeo—if anything— from the point of view of your own animal ethics."

Fifteen minutes later they came back. All took seats in the front, not the back. One man, the president of the club, stood nervously at the front of the room, hat in hand. "Well," I said, not knowing what to expect, nor what the change in attitude betokened, "what did you guys agree is wrong with rodeo?"

The president looked at me and quietly spoke: "Everything, Doc."

"Beg your pardon?" I said.

"Everything," he repeated. "When we started to think about it, we realized that what we do violates our own ethic about animals, namely, that you don't hurt an animal unless you must."

"Okay," I said, "I've done my job. I can go."

"Please don't go," he said. "We want to think this through. Rodeo means a lot to us. Will you help us think through how we can hold on to rodeo and yet not violate our ethic?"

To me this incident represents an archetypal example of successful ethical dialogue, using recollection, and judo rather than sumo.

This example has been drawn from an instance that involved people's personal ethics; the social ethic (and the law that mirrors it) has essentially hitherto ignored rodeo. But it is crucial to understand that the logic governing this particular case is precisely the same logic that governs changes in the social ethic as well. Here also, as Plato was aware, lasting change occurs by drawing out unnoticed implications of universally accepted ethical assumptions.

An excellent example of this point is provided by the civil rights movement in general, and more particularly by Martin Luther King Jr. and Lyndon Johnson's shepherding of the thinking and political activity that led to the monumental Civil Rights Act of 1964. As an astute politician, and particularly as an astute southern politician, Johnson had his finger on the pulse of how American segregationists were thinking. He realized that the social zeitgeist had progressed to the point that most Americans, even most southerners, accepted two fundamental premises, one ethical and one factual. The ethical assumption was that all humans should be treated equally in society, and the factual assumption was that blacks are humans. The problem was that many people had never bothered to put the two premises together and draw the inevitable conclusion, namely, that blacks should be treated equally. Johnson believed that if this simple deduction were put into law at that particular time, most people would "remember" and be prepared to bow to the inevitable conclusion. Had he been wrong, the Civil Rights Act would have been as meaningless as Prohibition, where a small subgroup of society attempted to force (sumo) its ethic on everyone else.

We have, in fact, over the last sixty years, lived through a good deal of Platonic ethical recollection regarding the ignored consequences of our accepted social ethic. We have seen that ethic rightfully extended not only to blacks but to women and other disenfranchised minorities when there was no morally relevant basis for withholding it. To deny an otherwise qualified woman admission into veterinary school, for example, on the grounds that she is a woman (a practice that was rife in these schools until the late 1970s) is as much a violation of the implications of our social ethic as is segregation. Nonetheless, getting people to recollect is a long, hard process, despite the simplicity of the argument on paper. But, still and all, social recollection has occurred, and we have become very much sensitized to remembering those groups of people hitherto disenfranchised and ignored.

The importance of judo—or recollection—cannot be overestimated. Too often we clash like linemen over ethical matters. We forget that, as remarked earlier, our ethical similarities, like our anatomical ones, are far greater than our differences. We are all brought up under the same laws and the same Judeo-Christian ethic; we watch the same movies and television programs, read the same newspapers and magazines, and share major portions of a culture. It is thus reasonable to assume that, if I detect something morally problematic, you will as well—*if* the problem is presented to you in such a way that you willingly, reflectively examine your own moral response rather than erect defenses. Thus social-ethical change as well as personal-ethical change proceeds optimally by recollection.

In fact, most of the ethical revolutions I alluded to as taking place over the last half century have also, like civil rights, depended on creating social recollection. In part because of the ingression of large numbers of women into the workforce during World War II as key players in the defense industry, society was better prepared than it had traditionally been to see women as humans protected by the Constitution and the Bill of Rights. A very similar realization helped drive the extending of protections to disabled Americans by the Americans with Disabilities Act. As a society, we began to realize that just because someone is confined to a wheelchair does not mean that they cannot function as a physician, or a lawyer, or a computer programmer, or, as in the famous World War II case of the legless wing commander Douglas Bader, who was strapped into a plane, a fighter pilot.

What does the foregoing have to do with creating a higher moral status for animals than they have historically enjoyed? If people do seek expansion of traditional limited ethics for animals, they are far more likely to look to our extant ethic for people than to generate a totally new ethic out of whole cloth. Although risking being called anthropomorphic, ordinary people who see animals confined in tiny cages will respond by saying, "How would you like to live under those kinds of conditions?"—implicitly applying ethical notions for people to animals. Obviously, much of the societal ethic will fail or fall short when exported to animals, but a good deal will not insofar as humans share a fair number of needs and desires with other animals, such as the need for security, food, water, companionship, stimulation, exercise, avoidance of pain, and myriad others. And so the fundamental question for anyone attempting to extend all or part of our social-ethical concerns to other creatures is this: Are there any morally relevant differences between people and animals that compel us to withhold the full range of our moral machinery from animals?

Answering this question occupied most of the thinkers who were trying to raise the moral status of animals during the 1970s and 1980s. While most philosophers working on this question did not affirm that there is no moral difference between the lives of animals and the lives of humans, there was a general consensus among them that *the treatment of animals by humans needs to be weighed and measured by the same moral standards by which we judge the moral treatment of humans.* Thus, for example, more and more people today judge severe agricultural confinement systems applied to animals more or less the way we would judge very severe confinement systems utilized to incarcerate humans: as constituting torture. If we are more comfortable with euthanizing suffering animals than euthanizing suffering people, it is very likely because we realize that death means something quite different to an animal than to a human. For animals, life exists more in the "now" than it does for people since animals generally lack future "projects," whereas such projects define "the meaning of life" for people, including things like "finishing my novel," "visiting Ireland one last time," and "seeing my grandchildren graduate college." The difference in future perspectives makes us far more hesitant to euthanize people to alleviate suffering, as does our poorly defined but widely believed notion that humans have free will while animals do not. It also makes us more willing to kill animals for food, provided the animals have had a good life. In the case of humans, Aristotle's dictum "Count no man happy until he is dead" seems more appropriate because we see human lives as far more complex than a series of "nows."

On the other hand, there are a considerable number of thinkers who have tried to deny a continuum of moral relevance across humans and animals and have presented arguments and criteria that support the concept of moral cleavage between the two. Many of these claims are theologically based. Most famous, perhaps, is the Catholic view that humans have immortal souls and animals do not, omnipresent across the Catholic tradition but with major exceptions that have been chronicled very deftly in the writings of Rod Preece. My own first book on the moral status of animals, *Animal Rights and Human Morality* (2006), originally published in 1982, devotes a good deal of attention to refuting such claims, including the notions that humans are more powerful than animals, are "superior" to animals, are higher on the evolutionary ladder than animals, are capable of reason and language while animals are not, are *moral agents* while animals are not, even that humans feel pain while animals do not. These fundamentally theologically based arguments draw a hard and fast line between humans, who have thoughts and feelings, and animals, who do not. (Regarding the claims about animals' lacking mind and the ability to feel pain, see my 1989 book *The Unheeded Cry: Animal Consciousness, Animal Pain, and Science.*) Mirabile dictu, it was not until the summer of 2012, at an international conference in Cambridge, England, that the scientific community stated that animals are conscious, a proposition by no means universally accepted by all scientific fields and individual scientists even today. We will shortly discuss the ideology in which the skeptical view of animal mind is embedded.

The Denial of Animal Mind

OF ALL THE arguments that various traditions have attempted to utilize to exclude animals from the moral arena, the most damaging are those stretching back to René Descartes that deny thought, feeling, and emotion to animals. It is obvious to ordinary common sense that we cannot have obligations to entities unless what we do to them, or allow to happen to them, *matters* to them. This is why we cannot have direct moral obligations to cars, diamond rings, golf clubs, or books. If I destroy a friend's car, I have not behaved in an immoral way toward the car, but only toward its owner, to whom the condition of the car matters. If I deface a valuable painting, I have not behaved immorally toward the painting, but rather toward the owner of the painting or toward myriad people whose joy in the painting has been summarily truncated. For this reason, anyone advocating for higher moral status for animals cannot let claims about lack of consciousness in animals go unchallenged and unrefuted.

In my experience most ordinary people, upon reflection, will acknowledge a continuum from animals through humans, as Charles Darwin did. Most people will affirm, when asked, that animals have thoughts and feelings. Even more important to the inclusion of animals within the scope of moral concern is the very Humean point that most people share empathetic identification with animals, particularly as regards their pain and suffering. (For David Hume, *sympathy* is the basis for ethics.)

Literally thousands of western cowboys and bikers have confessed to me their inability to watch movies, newsreels, or commercials depicting animal abuse and suffering—the ASPCA commercials showing abused and hurt animals with haunting music by Sarah McLachlan devastate some of the toughest, most macho men I know, who freely confess to needing to turn off the TV when these spots air. Any journalist who has covered stories of tragedies such as tornadoes and fires will tell you that stations get as much correspondence about and donations for the animals touched by the tragedies as for the people. A famous story concerns a young California mother killed on a jogging trail by a mountain lioness with cubs and the killing of the lioness by authorities because of the attack. More charitable contributions were received for the orphaned cubs than for the motherless children. A friend of mine who is a zoo veterinarian has told me a similar story about bear cubs orphaned by a tornado. One of my closest friends, a renowned psychiatrist, has told me of the impotent grief he felt when he stopped to help a dog hit by a car on a New York highway, and how he sat there weeping, knowing human medicine but not animal medicine and thence being unable to alleviate the animal's suffering. Let us recall countless firemen who return to burning buildings at the risk of life and limb to save pets. Let us also recall those people who have refused to be evacuated from life-threatening floods when authorities decreed that their animals must be left behind. One of the most touching TV stories coming out of the summer mountain wildfires in Colorado in 2013 showed a six-foot-four, 260-pound cowboy crying unabashedly when reunited with the horse he thought had been killed. Animal abusers, like child abusers, are shown little mercy by fellow criminals in prison.

Ordinary common sense, and by implication most ordinary people, has no problem with attributing mentation to animals—thoughts, feelings, emotions, intentions, pain, sadness, joy, fear, curiosity. If anything, most people exaggerate animal cognitive abilities. This uncritical imputation to animals of the full range of thought found in humans is certainly part of what led to the scientific community's calling any acknowledgment of mind in animals romantic anthropomorphism. But to argue that many people attribute far too much cognitive ability to animals certainly does not entail that one must deny *all mental states* to them. That the former occurs is indubitable. My wife worked with an educated colleague who was convinced that her dog knew when his birthday was coming and anticipated a special meal on that day. We can dismiss such a claim a priori on the grounds that, without a language, animals cannot have a concept of identifiable temporality.

We can also join with common sense in readily affirming that certain states are desirable to animals and others are abhorrent. As Hume pointed out, animals learn from a single instance of being stung by a bee that this is not an experience they wish to repeat. It is also noteworthy that the same David Hume who is famous for being one of the great skeptics in the history of philosophy, for doubting mind, body, causality, uniformity of nature, the knowability of nature, and the certainty of science, nonetheless found the existence of mind in animals indubitable.

In section 14 of his *Treatise of Human Nature,* "Of the Reason of Animals," Hume affirms that "next to the ridicule of denying an evident truth, is that of taking much pains to defend it; and no truth appears to me more evident, than that beasts are endowed with thought and reason as well as men. The arguments are in this case so obvious, that they never escape the most stupid and ignorant." (The last sentence is presumably directed at Descartes.)

It generally takes a PhD, MD, or DVM degree to evidence skepticism about animal mind. I remember entering a Harley-Davidson showroom after I had just published a new book. Some of the employees I knew there had cut out an article about my book and saved it for me. They asked me what the book was about. I replied that it was an attempt to prove to scientists and veterinarians that animals feel pain, and I was greeted with incredulity: "Who doesn't know that?" In other words, ordinary people are quite clear that animals experience the full range of feelings, positive and negative, that humans do.

When I first encountered skepticism about animals' feeling pain among veterinarians, I called a DVM PhD who had authored a chapter on felt pain in animals and modalities for mitigating and controlling such pain and asked him if he ever encounters veterinarians who deny the reality of felt pain in animals. "Oh, absolutely!" he replied. "How do you respond to such people?" I queried. In a heavy New York accent, he replied as follows: "I tell them to get a large male Rottweiler and place the dog on an examination table. I then tell them to take a vise grip, and fit it to the dog's testicles. I then tell them to *squeeze* the vise grip." He concluded that "the dog will tell you very clearly that it hurts *by ripping off your face!*"

In a classic experiment, the great Canadian psychologist David Hebb (Hebb, 1946) demonstrated that zookeepers could not do their jobs if they were prohibited from using mentalistic locutions about the animals in their care. And, as long as we are appealing to authority, we cannot forget that Darwin himself had no doubt about the presence of thoughts and feelings in animals.

For Darwin, the guiding assumption in psychology is continuity, so the study of mind is comparative, as epitomized by Darwin's marvelously blunt title for his 1872 work, *The Expression of the Emotions in Man and Animals*, a title that brazenly hoists a middle finger to the Cartesian tradition since Darwin saw emotion as inextricably bound up with subjective feelings. Furthermore, in *The Descent of Man* of the previous year, Darwin had specifically affirmed that "there is no fundamental difference between man and the higher mammals in their mental faculties" (1890, 66) and that "the lower animals, like man, manifestly feel pleasure and pain, happiness and misery" (1890, 69). In the same work, Darwin attributed the entire range of subjective experiences to animals, taking it for granted that one can gather data relevant to our knowledge of such experiences. Evolutionary theory demands that psychology, like anatomy, be comparative, for life is incremental and mind did not arise ex nihilo in man, fully formed, like Athena from the head of Zeus.

Darwin was of course not content to speculate about animal consciousness. He explicitly turned over much of his material on animal mentation to a trusted spokesman, George John Romanes, who in turn published two major volumes, *Animal Intelligence* (1882) and *Mental Evolution in Animals* (1883), both of which richly evidence phylogenetic continuity of mentation. In his preface to *Animal Intelligence*, Romanes acknowledges his debt to Darwin, who, in his words,

> not only assisted me in the most generous manner with his immense stores of information, as well as with his valuable judgment on sundry points of difficulty, but has also been kind enough to place at my disposal all the notes and clippings on animal intelligence which he has been collecting for the last forty years, together with the original manuscript of his wonderful chapter on "Instinct." This chapter, on being recast for the "Origin of Species," underwent so merciless an amount of compression that the original draft constitutes a rich store of hitherto unpublished material. (xi)

While Romanes focuses mainly on cognitive ability throughout the phylogenetic scale, he also addresses emotions and other aspects of mental life, all of which, for a Darwinian, ought to evidence some continuity across animal species.

In addition to the careful observations he made, Darwin also pursued a variety of experiments on animal mentation. He placed great emphasis on verifying any data subject to the slightest question. Toward this end, he, for example, contrived some ingenious experiments to test the intelligence of earthworms, a notion that he clearly felt was far beyond the purview of anecdotal information and that was sufficiently implausible as to require controlled experimentation. These experiments, now virtually forgotten, occupy some thirty-five pages of Darwin's book *The Formation of Vegetable Mould through the Action of Worms, with Observations on Their Habits,* published in 1881. The question Darwin asked was whether the behavior of worms in plugging up their burrows could be explained by instinct alone, by "inherited impulse" (1882, 67), or by chance, or whether something like intelligence was required. In a series of tests, Darwin supplied his worms with a variety of leaves, some indigenous to the country where the worms were found and others from plants growing thousands of miles away, as well as parts of leaves and triangles of paper, and observed how they proceeded to plug their burrows, whether using the narrow or the wide end of the object first. After quantitative evaluation of the results of these tests, Darwin concluded that worms possess rudimentary intelligence in that they showed plasticity in their behavior, some rudimentary "notion of . . . shape" (1882, 78, 99), and the ability to learn from experience. Darwin was no romantic anthropomorphist; he clearly distinguished the intelligence of the worms from the "senseless or purposeless" (1882, 97) manner in which even higher animals often behave, as when a beaver cuts up logs and drags them about when there is no water to dam or when a squirrel puts nuts on a wooden floor as if it had buried them in the ground.

We will shortly discuss how the belief that animals have thoughts and feelings, a belief that was sanctified by both common sense and Darwinian science, was nonetheless rejected by twentieth-century biology and psychology. Here it is sufficient to point out that ordinary common sense finds consciousness in animals to be essentially indubitable, though of course acknowledging that we can be wrong about animal mind in individual cases, as when people assume that a dog whose tail is wagging is in a friendly frame of mind.

It is worthwhile stressing at the outset that there are some cracks in the edifice of the denial of consciousness and mind in animals by the scientific ideology, or the ubiquitous dogma among scientists (which we will discuss in detail in chap. 6). As I have mentioned on numerous occasions, the general public has no trouble attributing thoughts, feelings, emotions, and all aspects of mentation to animals. In fact, historically it has been if anything excessive zeal in making such attributions to animals that has been dominant in the thought of ordinary people. Probably the mid-twentieth century represents the time when positivistic scientific ideology denying animal mind held the greatest stranglehold over science. However, a number of factors have further weakened the hold of this ideology in science as well as its influence in popular culture.

Probably most important culturally is the fact that the social mind has gradually grown more and more concerned about animal treatment and the ethics by which we judge such treatment. This is certainly the case regarding companion animals and has greatly ramified in legal mandates for pain control for research animals, proliferation of expensive veterinary specialties ranging from animal oncology to animal psychiatry, and willingness of the public to identify companion animals as "members of the family," as evidenced by surveys indicating that 80 to 100 percent of pet owners so view their animals.

Almost as important is the development of the Internet and specifically YouTube, which offers endless videos and stories evidencing animal emotion and mentation. A paradigm case of this is the story of Christian, a lion adopted by two Englishmen and eventually returned to his homeland in the wild. Years later, the adopters visited that homeland and recorded on film the lion's effusive reaction to their return. I could cite countless such videos, ranging from one of a dog apparently teaching a baby to crawl to footage of a male gorilla protecting a human child who had fallen into the animal's enclosure. The mainstream media quickly picked up on this trend, offering countless documentary programs dealing with animal behavior, animal friendships, animal thought, and so forth.

There is also a powerful minority of mainstream scientists who have rejected, implicitly or explicitly, the scientific ideology that denies animal mind. These people begin with Donald Griffin, who wrote the pioneering 1976 book defying behavioristic agnosticism or atheism regarding animal consciousness, *The Question of Animal Awareness.* Other pioneering figures include Jane Goodall, for her work on chimpanzees, and Marian Stamp Dawkins, for her revolutionary 1980 book *Animal Suffering: The Science of Animal Welfare.* And of course the work on teaching sign language to apes by various scientists, including R. A. and B. T. Gardner, Duane and Sue Rumbaugh, Roger Fouts, David Premack, and Penny Patterson, must similarly be credited. Other scientific work in this vein covers empathy in rats; reasoning and tool making, for example, in crows who are able to bend wire into a hook and insert it into a pipe to retrieve food, as was recorded in research by Alex Kacelnik and colleagues at the University of Oxford; and numerical concepts in various animals. We may also cite the example of Chaser, the border collie who learned over a thousand words for objects from her owner, the researcher John W. Pilley.

Such research, while vastly important, is the exception rather than the rule and is scattered randomly throughout the scientific literature. (The numerous works by Marc Bekoff on animal mind stand out as regular contributions to the literature.) In an anonymous review of Carl Safina's 2015 book *Beyond Words: What Animals Think and Feel* ("Reading Their Minds," 2015), an *Economist* writer is spot on, saying, in a very positive review, that the author is a "popularizer of science" and that in "the balance between science and storytelling" the author "favors the stories." In other words, the book is intended for the public, not for Safina's peers. While certainly serving to help erode the scientific ideology that denies scientific credibility to those who talk about animal mind, such talk is still not seen as "real science."

Even more important to our discussion, when scientists do affirm animal mentation, they rarely if ever connect it to issues of animal moral status, so at least that most unfortunate aspect of scientific ideology—that is, the concept of science as "ethics-free,"—is preserved. In addition, few science courses in universities address *either* animal mentation or animal moral status. Thus, I would submit that the scientific ideology that denies animal mind persists, albeit in a somewhat diminished way, even in the present moment.

Mattering and Telos

I WOULD ARGUE that the fact that many things *matter* to animals is absolutely obvious to common sense. When utilitarians like Bentham and Singer made the ability to experience pleasure and pain the sine qua non for being included in the moral arena, they were partly correct. Certainly, the ability to experience pleasure and pain represent *sufficient conditions* for moral status. That is to say, any being capable of experiencing pleasure and pain is a candidate for moral concern. But being able to feel pleasure and pain is not a *necessary condition* for moral status. Unless one stretches the notions of pleasure and pain so broadly as to include every possible positive or negative mental state that a being is capable of experiencing, thereby making these concepts essentially vacuous, a more appropriate concept is required. For example, if a person or animal is faced with something novel that could be dangerous, it is natural to say that that entity is *fearful*. Similarly, if either is placed in a greatly impoverished environment, it is natural to say that he or she is *bored*. It would be odd in either of these cases to say that the person or animal is in pain, even emotional pain. For this reason, I prefer to speak of things that *matter* to animals, for *mattering* far better captures the full range of positive and negative experiences that are morally relevant.

A moment's reflection will reveal that the kinds and cases of mattering are indefinitely large and varied. When, for example, my dogs observe me putting on shoes, they instantly grow very excited because they think I am taking them out to the significantly large acreage we live on. If I then say, "We're going out in the back" (i.e., the much smaller fenced backyard), they retain some excitement, but it is tempered by disappointment. If I come out of the barn with a saddle, some of my horses (i.e., those who like to be ridden) will demonstrate eager anticipation, while those who do not like riding will evidence evasion. These emotions reveal not only species-specific behavior but also marked individual differences. When Congress mandated environments for nonhuman primates that "enhance their psychological well-being," scientists quickly learned that there was no magic bullet. Some apes, for example, would obsessively focus on the game Simon (an electronic game wherein the player repeats sequences of flashing lights by pushing colored buttons in the same sequences), refusing to share with others, while some would show absolutely no interest in this pursuit. The lesson here is that one must attend not only to what an *animal's telos* dictates but also to individual variations and differences.

Obviously, telos, or animal nature, gives us a much richer category for evaluating *mattering* than does pleasure or pain. Shortly, we will evaluate the concept of telos in detail, in terms of its historical basis in Aristotle, its metaphysical status, and its acceptability to common sense. But here let me sketch it in broad brushstrokes.

I have thus far argued that the social-consensus ethic of the United States is based in the Constitution, the amendments thereto, and the laws derived therefrom. I have also argued that these laws change with the evolution of society, hopefully in the direction of greater justice. This ethic has built into it the mechanism for self-correction, as is evidenced by the changes introduced during the civil rights movement. I have explained that our social-ethical system is based partly in a utilitarian approach, attempting to produce the greatest benefit for the greatest number. This, however, runs the risk of having the majority ride roughshod over the interests of the minority. The notion of rights, protecting the individual and individual interests from being overwhelmed by the majority, provides a legally encoded protection from majoritarian excess. The interests that are protected, as enumerated in the Bill of Rights, represent those interests that are fundamental to human beings. These include freedom of religion, freedom of speech, protection against arbitrary search and seizure, due process and justice rendered by a jury of one's peers, freedom of the press, and protection against cruel and unusual punishment, and they are embodied in the Bill of Rights, which is basically a theory of human nature, that is, those interests that represent the most fundamental interests constitutive of human life, or human nature or human telos. If we are going to base animal ethics upon our social-consensus ethic, it stands to reason that we will likewise embody in law protection for *the fundamental interests of animals dictated by their natures, or teloi.*

In 1980, I gave a full-day seminar to representatives of all the Canadian ministry departments making policy for animal use, including Agriculture, Wildlife, Fisheries and Oceans, and so on. At the end of the seminar, the people present unanimously suggested that there was a need for a bill of rights for animals, to be determined by reference to their biological and psychological natures. Later that year, I received from an anonymous source at the Department of Fisheries and Oceans a copy of a reply by the minister to a request to take killer whales from Canadian waters in order to place them in the Vancouver Aquarium. The minister wrote that the request was denied until the aquarium could demonstrate that it had established accommodations for these animals that met their *telos-specific* needs. (In 2015 California law banned captive breeding of killer whales by SeaWorld.) This story powerfully evidences the degree to which the concept of telos appeals to the thought processes (or recollection) even of government officials, normally not known for flights of fancy.

One way to restate the latter point is that telos accords well with the *commonsense metaphysics of ordinary people.* Explication of this point requires an explanation of the preceding philosophical notion. A metaphysic is simply a set of concepts by which a person orders and categorizes the world. The scientific revolution of Descartes, Galileo, Newton, and others was at root a major change in what were accepted as the basic concepts needed to talk about reality. The basic metaphysical approach to reality before the scientific revolution was highly commonsensical and described by ordinary language. For people living then, the world consisted of what common sense and ordinary experience said it consisted of—flowers that were pretty and smelled good; fish that were alive and rocks that were not; entities that were hot and others that were cold. In short, the world was in reality exactly what common sense identified, a world of qualitative differences—beautiful and ugly; living and nonliving; hot and cold; good and evil. The scientists who created the scientific revolution saw the universe very differently. For them, colors, smells, tastes, and distinctions like living and dead were ultimately illusory. That is the point of Descartes's *Meditations,* that the world we access through our senses is illusory—what is real is what is described by mathematical equations. This is why Descartes takes such great pains to discredit what we know through the senses. If we could see the world the way that God does, he might say, we would see that it is simply qualitatively homogeneous matter configured in ways describable by geometrical physics. Thus, true knowledge comes from reason, not by way of the senses.

The scientific revolution that ushered in modern science, then, was rooted in revolutionary changes in *values*. Consider the development of modern physics at the hands of Descartes, Galileo, and Newton. While these thinkers did indeed lay the foundations for unprecedented changes in our empirical approaches to the world, a major part of their contribution involved pressing forward a new way of looking at reality, a new conceptual map of the furniture of the universe, or what one may rightly call a new metaphysic. While Aristotelian physics and its attendant view of the world stressed the need to explain the variety and irreducible qualitative differences found in experience—beautiful and ugly, living and nonliving, hot and cold—a view still echoed in ordinary common sense, the new physics explained away these qualitative differences as by-products of changes in fundamental, homogeneous, mathematically describable matter, the only legitimate object of study. Modern physics correlatively *disvalued* the sensory, the subjective, and the qualitative and, like Plato, equated the measurable with the real. Thus, whereas Aristotle saw biology as the master science and physics as, as it were, the biology of nonliving matter, Descartes saw biology as a subset of physics, and his thinking is therefore the herald of today's reductionistic molecular biology. And, of course, no amount of data gathered empirically can disprove Descartes's thesis, since what counts as data is determined by whether one buys the commonsense worldview or the reductionistic worldview of modern science. So modern science does not so much *disprove* the commonsense worldview as *disapprove of* it.

The metaphysics of modern science is barely four hundred years old. The metaphysics of Aristotle, on the other hand, lasted almost two thousand years. We can speculate that one reason it was so long-lived was its concordance with ordinary common sense. But the roots of the debate between commonsense metaphysics and a mathematically based, reductionistic metaphysics go back even further than Aristotle to the pre-Socratic philosophers. One influential group among the pre-Socratic *physikoi*, or physicists, as Aristotle called them, the thinkers attempting to explain reality and change, were the ancient atomists, among whom the seminal thinker was Democritus. The atomists foreshadowed modern science by affirming that all of reality is explicable by mechanical interactions between atoms, all qualitatively identical, moving in a void. Democritus enunciated the creed of mechanistic materialism very pithily: "By convention is sweet and sour, hot and cold. . . . In reality are only atoms and void" (Kirk and Raven, 1957). In other words, reality is made up only of atoms interacting in empty space. When the atoms constitutive of a lemon interact with the atoms making up the tongue and brain, they generate the experience of sourness. As John Locke later put it, sourness, warmth, and color are not *primary qualities*, that is, objective features of the real world, they are *secondary qualities*, subjective qualities arising from the interaction of the atoms making up objects and the atoms making up our sense organs and brain.

Aristotle had no tolerance for such a reductionistic approach. For him, reality is what we perceive—"what you see is what you get." Although a student of Plato's, he had no sympathy with Plato's notion that ultimate reality consists of abstract entities apprehended and understood by reason, the Forms. What is real for Aristotle is *tode ti*, what one can point to, "this here existent thing." Whereas mathematics for Plato represents the highest form of knowledge humans are capable of—eternal truths not subject to change—change, coming to be and passing away, is a primordial fact about the universe. In Aristotle's thinking, the truths of mathematics are trivial; they represent truths so abstract that they apply to everything.

For Plato, as for Descartes, mathematical physics obeying universal laws can and should be utilized to explain everything. Physics is the master science, and biology is simply applied physics. For Aristotle, biology is the master science, and physics is simply the biology of dead matter. Aristotle was a practicing biologist, describing and classifying living things in nature. He recognized what he called four causes, or explanatory principles, that answer four questions that can be asked about reality: What is it made of (the *material cause*)? What made it (the *efficient cause*)? What is its function or purpose (the *final cause*, or telos)? What is its nature (the *formal cause*)?

The mechanistic-materialistic thinkers put primary emphasis upon efficient causation coupled with material causation; think billiard balls. For them, there is one science of everything. For Aristotle, as for ordinary common sense, there are different ways of understanding different *kinds* of things—the behavior of birds is governed by different laws from the behavior of fish, snow, avalanches, falling water, rifles, plants. There is no single science of everything. All things are best explained in terms of functions and purposes. A human artifact, such as a knife, is best explained by its function—to cut, which is its telos. A living thing, be it mammal, bird, plant, insect, has a unique set of functions, making up its telos, or unique nature. As we said, living things are the paradigm for functional explanations. Every living thing is constituted of a set of functions making it a living thing and a unique living thing. All living things nourish, locomote, sense and perceive, excrete, and reproduce. How each living being actualizes those functions determines its telos. The science of biology is therefore the study of how each sort of living thing fulfills its "living," or way of life, or telos.

There is far more to life than pleasure and pain. All forms of "mattering" to an animal are determined by its telos. Some violations of telos ramify in conscious mattering. On the other hand, some living things do not register violations in consciousness. Plants require water and sunlight to actualize their teloi. But it is unlikely that they are aware of such mattering. Withholding water and sunlight from plants "matters" to them in the sense that they shrivel and die, albeit most likely without conscious awareness. Withholding water from an animal will also lead to the animal's shriveling and dying, but accompanied by negative feelings in animals capable of awareness and consciousness.

Animals like coyotes and raccoons, when caught in traps, will chew their legs off to escape. Clearly, that is more painful than being trapped. But the ability to escape, to be free, is more important to their telos than pain is, presumably because they are both predators and prey. This shows clearly that causing pain is not the worst thing one can do to an animal. (Compare the human hiker who cut his arm off with a pocket knife when it became wedged between a boulder and a rock wall, leaving him unable to escape.) These examples show that determining and prioritizing elements of telos is very much an empirical question.

We are assuming here, with ordinary common sense, that not all living things are capable of awareness. Some animals are clearly aware; plants are probably not aware, as awareness would do them no good, since they are rooted to a spot. The points along the phylogenetic scale where consciousness begins and ends are unclear, and can best be determined by empirical science, though without certainty. On the other hand, as we shall see, only a fool or an ideologically suffused scientist can doubt that higher animals possess some degree of awareness. Ordinary people, possessed of ordinary common sense, can no more deny that dogs and tigers and hawks are conscious than they can deny that their fellow humans are conscious. Thus the belief that many animals are conscious is solidly ensconced in the *metaphysics of ordinary common sense,* and, as I argued earlier, is indispensable in our ordinary-life interactions with and experiences of animals.

If we are to adopt telos as the basis for ethical obligations to animals, as our societal ethic has done for people, we can considerably broaden what is included in the moral arena or in the scope of moral concern. Since, from the traditional Aristotelian perspective, all natural objects possess teloi, and all artifacts made by humans do as well, we must differentiate between those teloi that demand moral concern from moral agents and those that do not. After all, in a teleological universe, even rocks have teloi, but it would be untenable, to say the least, to accord moral status to rocks. To answer this conundrum, we must refine the concept of *mattering.* It may well be the case that in a very restricted sense having oil matters to a car, since without it, the car will not run. But this is not a morally relevant sense of *mattering,* since the car neither knows nor cares whether it is running or not. *The kind of mattering that is morally relevant requires consciousness or awareness or caring in the entity in question.* For this reason, to enjoy moral status, an entity (i.e., an animal) must have the kind of telos whose violation *creates some negative mode of awareness in the creature in question.* As we have just discussed, ordinary people readily see many animals as possessing the requisite sort of consciousness. But members of the scientific community in the capacity of scientists do not readily accept the presence of consciousness in animals. We will later discuss why this is the case. But, in any event, such skepticism about animal mind serves to nullify concern about animal experience and therefore about animal moral status.

Thus, we are left with an interesting tension between a scientific worldview and a commonsense worldview. The former, based in mechanistic causation, has no room for consciousness—hence the mind-body problem that arises for Descartes. The latter is teleological and encompasses functions and values and thoughts and feelings. If one is a scientist, therefore, one is inexorably led away from subjectivity, ethics, and telos. Fortunately for animal ethics, the vast majority of ordinary people are not scientists, so both animal ethics and animal consciousness accord well with their take on reality or metaphysics. And thus our social ethic, melded with the concept of animal telos, gives us a way to expand the contents of the moral arena. Let us recall (with some bitterness), as mentioned earlier, that it was not until

2012 that the scientific community held a consensus conference at Cambridge University allowing that animals seem to have thought and feeling.

Specifically, when society confronts an animal issue not covered by the meager animal ethics currently available, for example, confining pregnant sows full time in tiny cages where they cannot turn around or even, in many cases, lie fully extended, it quickly becomes evident that such housing is totally incompatible with what we know of sow telos in nature. It then becomes necessary to socially "remind" producers (i.e., those who raise food animals for a living) of expectations regarding such housing. So much the better if producers voluntarily change. If they fail to do so, however, society can mandate proper housing that fits the needs emerging from swine telos. In a conceptual way, then, the sow has been given the *right* to housing fitting her biological and psychological nature. Since animals are, in the eyes of the law, human property, and since, legally speaking, property cannot have rights, the right to proper housing would take the form of how one must house one's "swine property." A legal right, resulting in a mandate, has been established, while circumventing the property issue. After all, we regularly restrict property use. For example, just because I own my motorcycle does not mean I can ride it on the sidewalk at a hundred miles per hour. In the animal case, we are restricting property use for moral reasons, specifically for animal ethics considerations.

Legal scholars across the United States are puzzling over the question of how rights can be directly granted to animals. That problem notwithstanding, we can now see how using our societal ethic for humans can function, *mutatis mutandis,* as a meaningful check against utilitarian excesses of animal use. The fact that this is very important to current society is graphically evidenced by the fact that, in 2004, *2,100 pieces of legislation designed to protect animal welfare were floated across the United States.*

We have reached a natural pause in our discussion, having outlined a basis for accomplishing what we set out to do, namely, to provide a theoretical structure for including animals and our treatment of them in the moral arena in a manner that gains acquiescence from ordinary people by appealing to what they already believe, that is, by reminding them of it.

To summarize: The societal ethic embodied in our social-consensus ethic provides a way of balancing utilitarian considerations with providing protection for essential aspects of human nature, or telos, as expressed in the Constitution. These protections, which build fences around individuals to prevent them from being oppressed by the majority for the general welfare, are called rights, and from them can be deduced additional rights, as our legal history demonstrates. Animals also have basic needs and interests constitutive of their nature, or telos, which we as a society believe should not be overridden for human good. As Aristotle pointed out, living beings have unique strategies for solving the problems inherent in living—sensing, moving, reproducing, nourishing themselves—the thwarting of which matters to them. The only way to avoid such a conclusion is to deny that certain things *matter* to animals, to affirm that they are nonconscious machines, a conclusion that is highly repugnant to ordinary people and common sense. The Aristotelian notion of telos thus becomes central to animal ethics.

Telos allows us to talk of animal interests, as we have seen, well beyond the narrow restrictions of pleasure and pain. In addition, it provides an excellent shield against the sorts of rationalizations that confinement agriculturalists in particular have given in defense of keeping animals in highly restrictive and impoverished environments. They tend to argue that in confinement systems, animals do not need to worry about finding food, extremes of climate, predators—all needs are supplied by the system. While there is certainly some truth in this argument, it neglects the fact that animals are built to, for example, find food in certain ways. Ethologists have shown that even domestic fowl, such as chickens, if given a choice of having ad libitum feed supplied to them rather than working for food, will choose to work.

A less than obvious example of violating animal nature may be found in a story recounted by Hal Markowitz, an expert on behavioral enrichment of environments for captive animals. Markowitz (Markowitz and Line, 1990) recounts that the Portland, Oregon, zoo built a showpiece exhibit for servals (a South African bobcat), even importing sand and plants from the Kalahari. The exhibit was a dud—the servals lay around in obvious depression, even refusing to eat. When Markowitz visited their native habitat, he found that the bulk of these animals' time was spent predating low-flying birds, their main source of food. He told the zoo that instead of feeding the servals horsemeat in chunks, the keepers should grind the rations into meatballs and shoot them randomly across the exhibit enclosure with a compressed-air cannon. The animals' behavior changed overnight—they became excited and active, clearly exercising the predating aspect of their telos. Despite the basic power of the food drive, it was trumped here by the failure to accommodate how the servals had evolved to eat.

Part Two

Ideology and Common Sense

Ideology and Common Sense

Ideology

BEFORE I DEMONSTRATE the value of the concept of telos to moving discussion of animal ethics forward, there are a number of points surrounding the concept that merit additional discussion. In particular, I described this book as being based in common sense. Yet our talk about differing and incompatible metaphysical views of the world that cannot be decided empirically and yet seem to be solidly grounded is indeed far removed from common sense. After all, it is a highly credible commonsense question to ask, Is the world the way our ordinary experience tells us it is, or is it what science tells us it is? And in a less commonsense vein, we might ask, Are there various "natural kinds" differing qualitatively in an irreducible way from each other, the way Aristotle claims and the way Descartes denies, or is it true that if we could see from a perspective of godlike accuracy, we would see a homogeneous reality differing in quantitative characteristics forming the illusion of qualitative differences? To use a homey analogy, are pancakes and coffee cake the same thing because they are made of the same ingredients (Bisquick, eggs, and water), or are they as different as our gustatory experiences tell us they are?

My inclination is to take a pragmatistic position on the question. Even in ordinary life there are very different answers to a given question, depending upon the way in which the question is asked, and all the answers may be true, but some may be totally inappropriate to the context of inquiry. Suppose, for example, I get a telephone call from a personnel officer of a large company to whom my name has been given as a reference by a former student, Ms. X, applying for a job. The company representative begins our conversation by asking me what I know of Ms. X. I respond by enthusiastically declaring that "she has the best forehand smash I have ever seen in any student." My statement may be absolutely true, but it is totally—and laughably—irrelevant given the context of our discussion. The interviewer may then respond by saying that he has no interest in her tennis prowess but rather in her ability to interact with clients. If I then proceed to say that "she will be fine in that kind of interaction as long as the client is not allergic to that odoriferous perfume she usually wears," there is little doubt that the interviewer will quickly thank me for my time and terminate the discussion, even though what I said may be true. It is not the truth or falsity of my assertion that is an issue—it is rather its relevance to the current context.

Similarly, in everyday experience we rarely ask about matters to which an answer from mathematical physics is the only plausible response. If you and I are discussing your lovely flower garden and I ask you how you manage to achieve such a wonderfully balanced array of colors, you would appear to be insane if you were to answer me in terms of the physics of color vision. Whether the world is "really" just a matter of homogeneous atoms randomly colliding in the void, that perspective on reality is rarely relevant to ordinary discourse. Insofar as our ordinary experience, which we cannot but treat as real, manifests differences between hot and cold, ugly and beautiful, good and bad, the reductionistic language of physics is inappropriate to ordinary conversation.

Nonetheless, the preceding point has been studiedly ignored by reductionistic scientists from Democritus to Descartes who are inclined to dismiss from their version of reality much of what we as human beings must deal with in our lives. It is precisely in keeping with this point that scientists have developed what I call the scientific ideology, or the common sense of science, for it is to science what ordinary common sense is to daily life.

What is an ideology? In simple terms, an ideology is a set of fundamental beliefs, commitments, value judgments, and principles that determine the way someone embracing those beliefs looks at the world, understands the world, and is directed to behave toward others in the world.

When we refer to a set of beliefs as an ideology, we usually mean that for the person or group entertaining those beliefs, nothing counts as a good reason for revising those beliefs and, correlatively, that raising questions critical of those beliefs is excluded dogmatically by the person with the belief system. (As twentieth-century political philosopher David Braybrooke has stated it, "Ideologies distort as much by omitting to question as by affirming answers" [1996, 126].)

The term is most famously, perhaps, associated with Karl Marx, who described capitalist ideology (or free-market ideology) as involving the unshakable beliefs that the laws of the competitive market are natural, universal, and impersonal; that private property in ownership of means of production is natural, permanent, and necessary; that workers are paid all they can be paid; and that surplus value should accrue to those who own the means of production.

Though most famously associated with the Marxist critique of capitalism, we all encounter ideologies on a regular basis. Most commonly, perhaps, we meet people infused with religious ideologies, such as biblical fundamentalism, who profess to believe literally in the Bible as the word of God. I have often countered such people by asking them if they have read the Bible in Hebrew and Greek, for surely God did not speak in antiquity in English. Further, I point out, if they have not read the original language, they are relying on interpretations rather than literal meaning since all translation is interpretation, interpretation that may be wrong. To illustrate this point, I ask them to name some of the Ten Commandments. Invariably, they say, "Thou shalt not kill." I then point out that the Hebrew in fact does not say, "Thou shalt not kill," it says, "Thou shalt not murder." This should be enough to convince them that they in fact do not believe the Bible literally, if only because they cannot read it literally. Does it do so? Of course not. They have endless ploys to avoid admitting that they cannot possibly believe it literally, for example, "The translators were divinely inspired," and so forth.

We of course are steeped in political ideology in grade school and high school, for example, on issues of "human equality." Ask the average college student (as I have done many times), What is the basis for professing equality, when people are clearly unequal in brains, talent, wealth, athletic ability, and so forth? Few will deny this, but most will continue to insist on "equality" without any notion that "equality" in our belief system refers to a way we believe we *ought* to treat people, not to a factual claim. If they do see equality as an "ought" claim, almost none can then provide a defense of why we believe we ought to treat people equally if in fact they are not equal. And so on. But virtually never will any student renounce the belief in equality.

Ideologies are attractive to people; they give pat answers to difficult questions. It is far easier to give an ingrained response than to think through each new situation. Militant Muslim ideology, for example, sees Western culture as inherently evil and corruptive of Islam and the United States as "the Great

Satan" and the fountainhead of Western culture, which in turn is aimed at destroying Islamic purity. The United States is thus automatically wrong in any dispute, and any measures are justified against the United States in the ultimate battle against defilement.

What is wrong with ideology, of course, is precisely that it truncates thought, providing simple answers and, as Braybrooke indicated in the statement quoted earlier, cutting off certain key questions. Intellectual subtlety and the powerful tool of reason in making distinctions are totally lost to gross oversimplification. Counterexamples are ignored. I recall working in a warehouse where the preponderance of blue-collar employees were strongly possessed of racist ideology, particularly antiblack ideology. It was universally believed that blacks are lazy, unintelligent, sneaky, and crooked. One day I had an inspiration. There was in fact one African American (Joe) who worked in the warehouse and was well liked. I raised this counterexample with some of the white workers. "Surely," I said, "this case refutes your claim about all black people." "Not at all," they said. "Joe is different—he hangs around with us!"

But it is not only that ideology constricts thought. It can also create monsters out of ordinary people by overriding common sense and common decency. We can see this manifested plainly throughout the history of the twentieth century. The recent experiences of Eastern Europe and Africa make manifest that ideologically based hatreds, whose origins have been obscured by the passage of time, may, like anthrax spores, reemerge as virulent and lethal as ever, unweakened by years of dormancy.

As we have seen, ideologies operate in many different areas—religious, political, sociological, economic, ethnic. Thus it is not surprising that an ideology would emerge with regard to science, which is, after all, the dominant way of knowing about the world in Western societies since the Renaissance. Indeed, knowing has had a special place in the world since antiquity. Among the pre-Socratics, or *physikoi* (physicists), one sometimes needed to subordinate one's life unquestioningly to the precepts of a society of knowers, as was the case with the Pythagoreans. And the very first line of Aristotle's *Metaphysics*—by which term he meant "first philosophy"—is "All men by nature desire to know." Thus the very telos of humanity, the "humanness" of humans, consists in exercising the cognitive functions that separate humans from all creation. Inevitably, the great knowers, such as Aristotle, Francis Bacon, Isaac Newton, and Albert Einstein, felt it necessary to articulate what separated legitimate empirical knowledge from spurious knowledge, and to jealously guard and defend the methodology used to distinguish between them from encroachment by false pretenders to knowledge.

Thus the ideology underlying modern (i.e., postmedieval) science has grown and evolved along with science itself. And a major—perhaps *the* major—component of that ideology is a strong positivistic tendency, still regnant today, of believing that real science must be based in experience since the tribunal of experience is the objective, universal judge of what is really happening in the world.

If one were to ask most working scientists what separates science from religion, speculative metaphysics, and shamanistic worldviews, they would unhesitatingly reply that it is an emphasis on validating all claims through sense experience, observation, or experimental manipulation. This component of scientific ideology tracks directly back to Isaac Newton, who proclaimed that he did not "feign hypotheses" ("hypotheses non fingo") but operated directly from experience. (The fact that Newton in fact did operate with nonobservable notions such as gravity, or, more generally, action at a distance, did not stop him from ideological proclamations affirming that one should not do so.) The Royal Society members apparently took him literally: they went around gathering data for their commonplace books and fully expected major scientific breakthroughs to emerge therefrom. (This idea of truth revealing itself through data gathering is prominent in Bacon.)

The insistence on experience as the bedrock for science continues from Newton to the twentieth century, where it reaches its most philosophical articulation in the reductive movement known as logical positivism, a movement that was designed to excise the unverifiable from science and, in some of its forms, to formally axiomatize science so that its derivation from observations was transparent. A classic and profound example of the purpose of the excisive dimension of positivism can be found in Einstein's rejection of Newton's concepts of absolute space and time, on the grounds that such talk was untestable. Other examples of positivist targets were Henri Bergson's (and some biologists') talk of life force (*élan vital*) as separating the living from the nonliving, and the embryologist Hans Driesch's postulation of "entelechies" to explain regeneration in starfish.

The positivist demand for empirical verification of all meaningful claims has been a mainstay of scientific ideology from the time of Einstein to the present. Insofar as scientists think at all in philosophical terms about what they are doing, they embrace the simple, but to them satisfying, positivism I have described. Through it, one can clearly, in good conscience, dismiss religious claims, metaphysical claims, and other speculative assertions not merely as false, and irrelevant to science, but as meaningless. Only what can be in principle verified (or falsified) empirically is meaningful. "In principle" means "someday," given technological progress. Thus, though the statement "There are intelligent inhabitants on Mars" could not in fact be verified or falsified in 1940, it was still meaningful since people could see how it could be verified, that is, by perfecting rocket ships and going to Mars to look. Such a statement stands in sharp contradiction to the statement "There are intelligent beings in heaven" because, however our technology is perfected, we cannot visit heaven, it not being a physical place. (Ironically, the emphasis on empirical verification clashes with the belief that the world is really what physics tells us.)

What does all this have to do with ethics? Quite a bit, it turns out. Ethics is not part of the furniture of the scientific universe. You cannot, in principle, test the proposition that "killing is wrong." It can be neither verified nor falsified. So, empirically and scientifically, ethical judgments are meaningless. From this, it was concluded that ethics is outside of the scope of science, as are all judgments regarding values rather than facts. The concept that I in fact learned in my science courses in the 1960s and that has persisted to the present is that science is "value-free" in general and "ethics-free" in particular. This denial in particular of the relevance of ethics to science is explicitly stated in science textbooks.

In addition to being explicitly affirmed, this component of the scientific ideology has also been implicitly taught in countless ways. For example, student moral compunctions about killing or hurting an animal, whether in secondary school, college, graduate school, or professional school, were never seriously addressed until the mid- to late 1980s, when the legal system began to entertain conscientious objections. One colleague of mine, who was in graduate school in the late 1950s studying experimental psychology, tells of being taught to "euthanize" rats after experiments by swinging them around and dashing their heads on the edge of a bench to break their necks. When he objected to this practice, he was darkly told, "Perhaps you are not suited to be a psychologist." In 1980, when I began to teach in a veterinary school, I learned that the first laboratory exercise required of the students, in the third week of their first year, was to feed cream to a cat and then, using ketamine (which is not an effective analgesic for visceral pain but instead serves to restrain the animal), do exploratory abdominal surgery ostensibly to see the transport of the cream through the intestinal villi. When I asked the teacher the point of this horrifying experience (the animals cried out and showed other signs of pain), he told me that it was designed to "teach the students that they are in veterinary school, and needed to be tough, and that if they were 'soft,' to 'get the hell out early.'"

As late as the mid-1980s, most veterinary and human medical schools required that the students participate in bleeding out a dog until it died of hemorrhagic shock. Although Colorado State University's veterinary school abolished the lab in the early 1980s for ethical reasons, the department head who abolished it was defending the practice ten years later, after moving to another university, and explained to me that if he did not, his faculty would force him out. As late as the mid-1990s, a medical school official told the veterinary dean at my institution that his faculty was "firmly convinced" that one could not "be a good physician unless one first killed a dog." In the autobiographical book *Gentle Vengeance*, which deals with an older student going through Harvard Medical School, the author remarks in passing that the only purpose he and his peers could see to the dog labs was to assure the students' divestiture of any shred of compassion that might have survived their premedical studies.

Veterinary surgery teaching well into the 1980s was also designed to suppress compassionate and moral impulses. In most veterinary schools, animals (most often dogs) were operated on repeatedly, from a minimum of eight successive surgeries over two weeks to over twenty times at some institutions. This was done to save money on animals, and the ethical dimensions of the practice were never discussed, nor did the students dare raise them.

At one veterinary school, a senior class provided each student with a dog, and the student was required to do a whole semester of surgery on the animal. One student anesthetized the animal, beat on it randomly with a sledge hammer, and spent the semester repairing the damage. He received an A.

The point is that these labs in part taught students not to raise ethical questions and that ordinary ethical concerns were to be shunted aside, and ignored, in a scientific or medical context. So the explicit denial of ethics in science was buttressed and taught implicitly in practice. If one did raise ethical questions, they were met with threats or a curt "This is not a matter of ethics but of scientific necessity," a point that was repeated when discussing questionable research on human beings.

Even at the height of concern about animal use in the 1980s, scientific journals and conferences did not rationally engage the ethical issues occasioned by animal research. It was as if such issues, however much a matter of social concern, were invisible to scientists, which in a real sense they in fact were. One striking example is provided by a speech given by James Wyngaarden, the director of the National Institutes of Health, in 1989. The NIH director is arguably the chief biomedical scientist in the United States and certainly is a symbol of the research establishment. Wyngaarden, an alumnus of Michigan State University, was speaking to a student group at his alma mater and was asked about ethical issues occasioned by genetic engineering. His response was astonishing to laypeople, though perfectly understandable given what we have discussed about scientific ideology, or the common sense of science. He opined that, while new areas of science are always controversial, "science should not be hampered by ethical considerations" (Michigan State University, 1989). Probably no other single incident shows as clearly the denial of ethics in science. When I read the unattributed quotation to my students and ask them to guess its author, they invariably respond, "Adolf Hitler."

Nor is this sort of response restricted to biomedicine. Some years ago, PBS ran a documentary special on the Manhattan Project, which developed the atomic bomb. Scientists on the project were asked about the ethical dimensions of their work. They replied that the ethics was not their business; society makes ethical decisions, scientists simply provide technical expertise regarding the implementation of those decisions. In fact, every time I am interviewed by a reporter on ethical issues in science, my raising the "science is value-free" component of scientific ideology elicits a shock of recognition. "Oh yeah," they say, "scientists always say that when we ask them about controversial issues like weapons development."

It is therefore not surprising that when scientists are drawn into social discussions of ethical issues they are every bit as emotional as their untutored opponents. It is because their ideology dictates that these issues are nothing but emotional, that the notion of rational ethics is an oxymoron, and that he who generates the most effective emotional response "wins."

Just how extraordinarily incapable scientists are of responding to rational ethical argument was driven home to me when I ran a long session on animal ethics and legislation at a 1982 national meeting of the American Association for the Advancement of Laboratory Animal Science (AAALAS), where I carefully laid out the arguments for legislating protections for research animals. Though the audience of laboratory-animal veterinarians expressed great frustration that researchers did not listen to them, particularly in human medical schools, and that their expertise, if attended to, would make for better animal care *and* better science, they steadfastly refused to support their own legislative empowerment since they opposed the importation of ethics into science!

As irrational as that was, it paled in comparison to what occurred after my session. Reporters converged on the president of the AAALAS, asking him to comment on my demand for legislated protection for animals. "Oh, that is clearly wrong," he said. "Why?" they queried. "Because God said we could do whatever we wish with animals," he affirmed. The reporters then turned to me and asked me to respond. Amazed that the head of a scientific organization could so invoke the Deity with a straight face (imagine the head of the American Physical Society responding to budget cuts in the funding of physics by saying, "God said we must fund physics"), I poked fun at his reply. "I doubt he is correct," I answered. "He comes from Kansas State University." "So what?" said the reporters. "Simple," I replied. "If God chose to reveal his will at a veterinary school, it certainly would not be at Kansas State! It would be at Colorado State, which is God's country!"

What are we to say of the aspect of scientific ideology that denies the relevance of values in general and ethics in particular to science? As I hope the astute reader has begun to realize, as a human activity, embedded in a context of culture, and addressed to real human problems, science cannot possibly be value-free, or even ethics-free. As soon as scientists affirm that controlled experiments are a better source of knowledge than anecdotes, that double-blind clinical trials provide better proof of hypotheses than asking the Magic 8 Ball, or, for that matter, that science is a better route to knowledge of reality than mysticism, we encounter value judgments as presuppositional to science. To be sure, they are not all ethical value judgments but rather epistemic ("pertaining to knowing") ones, but they are still enough to show that science does depend on value judgments. So choice of scientific method or approach represents a matter of value. Scientists often forget this obvious point; as one scientist said to me, "We don't make value judgments in science; all we care about is knowledge."

In fact, reflection on the epistemic basis of science quickly leads to the conclusion that this basis includes moral judgments as well. Most biomedical scientists will affirm that contemporary biomedicine is logically (or at least practically) dependent on the use of—sometimes the invasive use of—animals as the only way to find out about fundamental biological processes. Every time one uses an animal in an invasive way, however, one is making an implicit moral decision, namely, that the information gained in such use morally outweighs the pain, suffering, distress, or death imposed on such an animal to gain the knowledge or that it is morally correct to garner such knowledge despite the harm done to animals. Obviously, most scientists would acquiesce to that claim, but that is irrelevant to the fact that it is still a moral claim.

Exactly the same point holds regarding the use of human beings in research. Clearly, unwanted children or disenfranchised and marginalized humans are far better (i.e., higher-fidelity) experimental models for the rest of us than are the typical animal models, usually rodents. We do not, however, allow unrestricted invasive use of humans despite their scientific superiority. Thus another moral judgment is presuppositional to biomedical science.

I was once arguing with a scientist colleague about the presence of moral judgments in science. He was arguing their absence. I invoked the argument that, if science were ethics-free, we would always use the highest-fidelity model in our researches, thus deploying unwanted children rather than rats. In the ensuing silence, I asked him again: "Why not use the children?" "Because they won't let us!" he snapped.

In any case, many other valuational and ethical judgments appear in science, not just those involved in methodology. Which subjects and problems scientists are funded to pursue—AIDS, nonpolluting energy sources, alcoholism, but not the tensile strength of blonde hair or the intelligence of frogs—depends on social value judgments, including ethical ones. (Engaging in scientific research today depends on funding from federal agencies or private enterprise.) The once popular scientific subject of race, or the measurement of an alleged biological property called IQ, are now forbidden subjects for ethical reasons, as are myriad other subjects inimical to current social-ethical dogmas and trends.

Even experimental design in science is constrained by ethical value judgments. The statistical design of an experiment testing the safety of a human drug will invariably deploy far greater statistical stringency than a similar experiment testing the safety of an animal drug used for precisely the same disease in animals, for ethical reasons of valuing harm to people as a much greater moral concern than harm to animals.

The root paired concepts of biomedical science—health and disease—can also be readily shown to contain irrevocably valuational components. Physicians are convinced that the judgment that something is diseased or sick is as much a matter of fact as is the judgment that the organism is bigger or smaller than a breadbox. Diseases are repeatable entities to be scientifically discovered—physicians are scientists. This scientific stance has been repeatedly noted in its nonsubtle manifestations; anyone who has been in a hospital is aware of the tendency of physicians to see patients as instances of a disease rather than as unique individuals—science after all deals with the repeatable and law-like aspects of things, not with individuals qua individuals. This tendency to remove individuality is a chronic complaint of patients—it is demeaning to be treated as an instance of something. Indeed, it is less often noticed that this tendency is medically pernicious as well. When it comes to dispensing pain medication, for example, it has been shown that pain tolerance thresholds (i.e., the maximum pain a person can tolerate) differ dramatically across individuals and that thresholds can be modulated by a variety of factors, not the least of which is surely rapport with the physician, or the sense that the physician cares about the patient's pain. Among physician authors, only Oliver Sacks, in *Awakenings,* has stressed the extraordinary degree to which a disease varies with the individuals, in all their complexity, in whom that disease is manifested or instantiated.

This much ordinary common sense (but not the common sense of science) recognizes. The more subtle sense in which scientism in its emphasis on fact versus value—with only the former term entering into the medical situation—misses the mark is in its understanding of the very nature of disease. For the concept of a disease, of a physical (or mental) condition in need of fixing, is inextricably bound up with valuational presuppositions. Consider the obvious fact that the concept of disease is a concept that, like good and bad, light and dark, acquires its meaning by contrast with its complement, in this case, the concept of health. One cannot have a concept of disease without at least implicit reference to the concept of health (that is, okay and not in need of fixing). Yet the concept of health clearly makes tacit or explicit reference to an ideal for the person or other organism; a healthy person is one who is functioning as we believe people should. This ideal is clearly valuational; most of us do not feel that people are healthy if they are in constant pain, even though they can eat, sleep, reproduce, and so forth. That is because our ideal for a human life is really an ideal for a *good* human life—in all its complexity.

Health is not merely what is statistically normal in a population (statistical normalcy can entail being diseased); nor is it purely a biological matter. The World Health Organization captures this idea in its famous definition of health as "a complete state of mental, physical, and social well-being." In other words, health is not just of the body. Indeed, the valuational dimension is both explicit and not well defined, for what is "well-being" save a value notation to be made explicit in a sociocultural context?

Heedless of this point, and wedded to the notion that disease is discovered by reference to facts, not in part decided by reference to values, physicians make decisions that they think are discoveries. When physicians announce that obesity is the number one disease in the United States, and this "discovery" makes the cover of *Time* magazine, few people, physicians or otherwise, analyze the deep structure of that statement. Are fat people really sick people? Why? Presumably the physicians who make this claim are thinking of something like this: fat people tend to get sick more often—flat feet, strokes, bad backs, heart conditions. But, one might say, is something that makes you sick itself a sickness? Boxing may lead to sinus problems and Parkinson's disease—that does not make it in itself a disease. Not all or even most things that cause disease are diseases.

Perhaps the physicians are thinking that obesity shortens life, as actuarial tables indicate, and that is why it should be considered a disease. In addition to being vulnerable to the previous objection, this claim raises a more subtle problem. Even if obesity does shorten life, does it follow that it ought to be corrected? Physicians, as is well known, see their mission (their primary value) as preserving life. Others, nonphysicians, however, may value quality over quantity for life. Thus, even if I am informed—nay, guaranteed—that I will live 3.2 months longer if I lose forty-five pounds, it is perfectly reasonable for me to say that I would rather live 3.2 months less and continue to pig out. In other words, to define obesity as a disease is to presuppose a highly debatable valuational judgment.

Similar arguments can be advanced vis-à-vis alcoholism or gambling or child abuse as diseases. The fact that there may be (or are) physiological mechanisms in some people predisposing them to addiction does not in and of itself license the assertion that alcoholics (or gamblers) are sick. There are presumably physiological mechanisms underlying all human actions—flying off the handle, for example. Shall physicians then appropriate the management of temperament as their purview? (They have, in fact.) More to the point, shall we view people quick to anger as diseased—Doberman's syndrome?

Perhaps. Perhaps people would be happier if the categories of badness and weakness were replaced with medical categories. Physicians often argue that when alcoholism or gambling are viewed as sickness, that is, something that happens to you that you cannot help, rather than as something wrong that you do, the alcoholic or gambler is more likely to seek help, knowing he or she will not be blamed. I, personally, am not ready to abandon moral categories for medical ones, as some psychiatrists have suggested. And, as Kant said, we must act as if we are free and responsible for our actions, whatever the ultimate metaphysical status of freedom and determinism. I do not believe that one is compelled to drink by one's physiological substratum, though one may be more tempted than another with a different substratum.

I recall one occasion where one of my freshman students came to my office hours, visibly upset and indeed on the verge of tears. When I asked him what the problem was, he told me that he had visited a physician for a routine checkup. In the course of taking his history, the physician determined that both his parents were alcoholics and told the student that he could not go to venues where alcohol was consumed, and specifically cited student parties. The student was very upset because this advice dealt a serious blow to his social life, and he asked my opinion. I told him that I thought he could go to parties and other such events, provided he remained very aware of the need not to drink. "Take a glass of ginger ale or Coke, consume it slowly so no one attends to what you're drinking. Take no alcohol and act normally. Keep me posted on your progress." Three years later, when he was about to graduate, he came back to see me and thanked me warmly for the advice. Following it, he had been able to enjoy a normal social life and had not become an alcoholic even in the presence of alcohol. He had been scrupulously careful not to consume alcohol, and was in fact never tempted to do so.

Be that as it may, the key point is that physicians are not *discovering* in nature that conditions like obesity or alcoholism are diseases, though they think they are. They are, in fact, promulgating values as facts and using their authority as experts in medicine to insulate their value judgments from social debate. This occurs because they do not see that facts and values blend here. They are not ill intentioned, but they are muddled, as is society in general. And to rectify this, we must discuss, in a democratic fashion, which values will underlie what we count as health and disease, not simply accept value judgments from authorities who are not even cognizant of their existence, let alone conceptually prepared to defend them. At the very least, if we cannot engender a social consensus, we should articulate these for ourselves.

In 1988, the Environmental Protection Agency rejected scientifically sound toxicological data on moral grounds because the experiments that generated it were done by the Nazis on human beings against their will, out of fear of legitimating such experimentation. This decision was made despite the fact that other well-established areas of science—such as research on hypothermia and human reactions to high altitude beginning in the 1940s—had been based on and derived from Nazi experiments, and despite the fact that failing to use the data essentially entailed that much invasive animal research needed to be done to replace it.

Consider a revolution that I have looked at in considerable detail in another book, the replacement of psychology as the science of consciousness by behaviorism, which saw psychology as the science of overt behavior and ignored internal mental states. What facts could force such a change? After all, we are all familiar with the existence of our subjective experiences. Few people were impressed with behaviorism's founder John B. Watson's denial of consciousness (he came perilously close to saying, "We don't have thoughts, we only think we do"). Rather, people were moved by his valuational claims that studying behavior is more valuable because we can learn to control it.

Clearly, then, the component of scientific ideology that affirms that science is value-free and ethics-free is incorrect. We can also see that the more fundamental claim—that science rests only on facts and includes only what is testable—is also badly wrong. How, for example, can one scientifically prove (that is, empirically test) that only the verifiable may be admitted into science? How can we reconcile the claim that science reveals truth about a public, objective, intersubjective world with the claim that access to that world is only through inherently private perceptions? How can we know that others perceive as we do, or, indeed, perceive at all, since we cannot verify the claim that there are other subjects? How can science postulate an event at the beginning of the universe (the Big Bang) that is by definition nonrepeatable, nontestable, and a singularity? How can we know scientifically that there is reality independent of perception? How can we know scientifically that the world was not created three seconds ago, complete with fossils and us with all our memories? How can we verify any judgments about history? How can we reply that we know things best when we reduce them to mathematical physics rather than when we stay at the level of sensory qualities? Answers to the above questions are not verified scientifically. In fact such answers are presuppositional to scientific activity.

I have in fact alluded to another component of scientific ideology that worked synergistically with the denial of values to remove animal ethics from the purview of science. This is the claim that we cannot legitimately speak of thoughts, feelings, and other mental states in science since we cannot deal with these things objectively, not having access to the thoughts and consciousness of others. As I have explained elsewhere, this denial allowed scientists to negate the reality of animal pain, distress, and fear, *while at the same time using animals as models to study pain.* In a previous book, *The Unheeded Cry,* I demonstrated that this viewpoint was adopted in the early twentieth century by behavioral psychologists despite the fact that the dominant approach to biology was Darwinian, and Darwin himself, and most of his followers, eloquently affirmed that if morphological and physiological traits are phylogenetically continuous, so too are mental ones. I showed in that book, I hope, that the removal of thought and feeling from legitimate science was not a matter of new data that refuted old attempts to study animal mind, nor was it a result of someone's finding a conceptual flaw in that old approach (as Einstein did with Newton's views of absolute space and time). In fact, the shift to studying behavior rather than mind was effected by valuational rhetoric, namely, that if we study behavior rather than thought, we can learn to shape it and modify it—to extract behavioral technology from science, as it were. Anyway, the rhetoric continued, real sciences like physics deal with observables (a claim not always true—consider particle physics), and if we want to be real scientists, we need to lose subjectivity. So despite the ideological belief that science only changes by empirical or logical falsification, we have shown that, at least in psychology, a major change in what counted as scientific legitimacy was driven by values.

Another component of scientific ideology that follows closely upon our discussion of values is the ubiquitous belief that we best understand any phenomenon when we have understood it at the level of physics and chemistry, ideally physics. It is this component of scientific ideology that led a very prominent colleague of mine in physiology who works on fascinating issues in animal evolution at the phenotypic level to affirm in one of my classes, "Science has passed me by. . . . My work is archaic. . . . All real science now operates at the molecular level."

This reductionistic approach further removes scientists from consideration of ethics. If what is "really real" and "really true" is what is described by physics, it is that much easier to treat ethical questions arising at the level of organisms as being as "unreal" or "untrue" as the level at which they arise. The language of physics is, after all, mathematics; yet ethical questions seem inexpressible in mathematical terms. The belief that expressing things mathematically, as physics does, is getting closer to the truth leads in fact to a kind of "mathematics envy" among areas of science less quantitative, and sometimes leads to pseudo-mathematical obfuscation being deployed in fields like sociology or psychology to make these fields appear closer to the reductionistic ideal. In the end, of course, as I pointed out regarding the scientific revolution, a commitment to reductionism represents a value judgment, not the discovery of new facts. No empirical facts force the rejection of qualitative work for quantitative, and Aristotle, for one, explicitly rejected such rejections.

The final elements of scientific ideology worth mentioning are the beliefs that science should be ahistorical and aphilosophical. If the history of science is simply a matter of "truer" theories replacing false or partially false ones, after all, why study a history of superseded error? How things come to be accepted, rejected, or perpetuated is ultimately seen as not being a scientific question. Thus many scientists lack a grasp of the way in which cultural factors, values, and even ethics shape the acceptance and rejection of whole fields of study (for example, consciousness, as we have already discussed, eugenics, intelligence, race, psychiatry as a medical discipline, and so on). To take one very interesting example, it has been argued that quantum physics in its current form would never have been possible without the cultural context prevailing in Germany between 1918 and 1927.

Historian Paul Forman (1971) has argued that a major impetus for both the development and the acceptance of quantum theory was a desire on the part of German physicists to adjust their values and their science in the light of the crescendo of indeterminism, existentialism, neoromanticism, irrationalism, and free-will affirmation that pervaded the intellectual life of Weimar and that was hostile to deterministic, rationalistic physics. Thus quantum physicists were enabled to shake the powerful ideology of rationalistic, deterministic, positivistic late nineteenth- and early twentieth-century science, with its insistence on causality, order, and predictability, as a result of the powerful social and cultural ambience in German society that militated in favor of a world in which freedom, randomness, and disorder were operative and that valued such chaos both epistemically and morally.

The rejection of philosophical self-examination is also built into scientific ideology and into scientific practice. Since philosophy is not an empirical discipline, it is excluded by definition by scientific ideology from the ken of science. Further, historically, philosophy, like theology, competed with science, at least in the area of speculative metaphysics, so the few historically minded scientists approached it with suspicion, which spread to others. In any case, scientists do not have time for "navel-gazing" or "pontificating," as they often characterize philosophy—they are too busy doing science to reflect much on it. As one scientist said to me, "When I win a Nobel Prize, then I will write philosophy, because then everyone will want to read it, whether it makes sense or not!" Clearly, then, reflection on science and ethics must also await a Nobel Prize.

The reader will note that many of my examples of scientific ideology have been drawn from earlier decades. This is because scientific ideology was largely uncritically accepted in the public mind because the public mind was unconcerned with what scientists believed—scientists believed many strange things—so scientists thought little about making public statements that revealed various aspects of their thinking, such as when the scientific community looked askance at the work of Jane Goodall, or failed to respect and share the public's concern for ethical issues raised by science, in areas from genetic engineering to animal research. With the public now more aware of what scientists think, scientists have become more guarded, for example, regarding the denial of animal consciousness. Blatant disregard for social and ethical concerns put scientific freedom at risk, and science became more circumspect, even though the basic ideology was unchanged.

An excellent example is provided by the fact that ethical issues are still neither taught nor discussed in science courses. However, in 1990 the NIH mandated the teaching of what the government considers research ethics, instruction in "the responsible conduct of research," concerning regulatory compliance, wherein a series of "thou shalt not" policies are taught like a religious catechism in a few days, with no discussion, by people with no ethics background whatever. (In one case, a chemistry professor at my university attended a *two-day* seminar in teaching ethics and earned a "certificate" designating her as a "qualified ethics instructor." She had never had a philosophy class, but what the hell . . . it was just ethics! I asked an artist friend of mine to make a plaque for me certifying that I had had more than two days of chemistry instruction and thus was qualified to teach quantum chemistry.)

Such contempt for ethics teaching pervades virtually all such classes. As it happens, I have taught a number of science classes at an upper level as well as having taught ethics for more than thirty years, and I can unequivocally affirm that it is much harder to teach ethics properly. Recently, I had numerous students in my graduate course Science and Ethics who had taken the "regulatory compliance" class and, in writing, pronounced the difference between the two courses as "night and day." As one student wrote, "Regulatory compliance taught us a list of what to do and what not to do—you taught us *why!*"

Now that we understand ideology and the way it operates in the common sense of science, we can return to discussing the way multiple metaphysics can exist in the same individual. Consider an analogy: Imagine a person who requires glasses to see normally at his workplace. The only problem is that, due to the peculiarity of his optical correction, these glasses do not allow him to see anything colored purple as purple; instead, he sees it as gray. Eventually, he may forget that the work world contains purple items. Similarly, if one is sufficiently indoctrinated with scientific ideology, and has grown accustomed to working with peers who share that ideology, *awareness of ethical issues in science simply does not come up in one's thinking or conversation.* If one interacts with nonscientists who do see ethical issues in science, they can simply be dismissed as lacking a scientific perspective, in the same way that an educated person dismisses scientifically naive people who see Australians as living "upside down."

One important distinction must be drawn here. Some ideologies typically pervade every aspect of one's life. Racism is this sort of ideology. Scientific ideology tends to be restricted to the context of one's scientific activity. For example, John B. Watson was ideologically committed to denying consciousness to other beings, human or animal. One can, however, be morally certain that when Watson went home to his wife and she began a conversation about some newsworthy issue by saying, "Do you know what I think, John?" he of course did not respond by saying, "I doubt that you think at all." In other words, one's ideological commitments are checked by the pressures of ordinary life. The pressure of donning and doffing ideology creates a condition in people that psychologists call compartmentalization. When I was an undergraduate at the City College of New York, I was aware of a group called the Society of Orthodox Jewish Scientists. Fundamentalist scientists exist everywhere and are quite capable of believing that the world is five thousand years old, while also fully accepting an age of billions of years in their scientist moments. Many scientists, Descartes included, treat animals with doting and affection in their ordinary lives (he raised spaniels and gave them as gifts), while seeing them as machines incapable of pain in the course of their scientific activities. In the same way, many scientists espouse politically liberal causes in their ordinary lives, while concurrently denying any sense to ethical judgments. Of paramount importance is the fact that coexisting ideologies rarely confront one another, hence the pervasiveness of compartmentalization.

As important as the denial of ethics is to scientific ideology, equally important is the denial of consciousness, thought, feeling or any other mode of awareness and evidence of mental activity. In fact, as we saw earlier, the absence of mind in animals, so that nothing matters to them, would constitute a morally relevant difference removing them from the scope of moral concern. Recall that Descartes conveniently made the attribution of consciousness dependent on possession of language. As we shall see, there are indeed certain aspects of mind that do require language. Without language one cannot think in negative terms (there are no jaguars in the library), fictional terms (Superman grew up in Smallville), counterfactual terms (if it doesn't rain we can meet in the park), or futural terms (I want to visit Ireland next summer). But one can certainly experience the negative thoughts and feelings associated with violation of telos, as well as the pleasures and satisfaction that come with meeting telos demands. The important philosophical question that arises is this: It appears that attributing mental states to animals requires that we utilize anthropomorphic inferences. All sorts of scholars, particularly scientists in the biological and psychological sciences, see anthropomorphism as a cardinal sin. Can one construct an argument that makes anthropomorphism legitimate? Before embarking on that question, we need to remind ourselves that telos violations can be so egregious that one need not be Jane Goodall to be cognizant of them—witness sow stalls, giraffe and orca enclosures, veal crates, and so on—this is why ordinary people not grounded in ethology are so shocked by them.

Anecdote, Anthropomorphism, and Animal Mind

IT IS VIRTUALLY impossible to emerge from a training program in the biological or biomedical sciences without having developed a well-honed skepticism about and distaste for anthropomorphic attribution of mental states to animals. Equally suspect to the biology graduate is the attempt to evidence such states by appeal to anecdotal information of the sort routinely accepted by ordinary common sense.

The inculcation of this skepticism into nascent scientists is a major part of what I call the common sense of science, or scientific ideology, the set of foundational or philosophical assumptions taught as fact along with the empirical material constitutive of the scientific discipline in question. In the case of animal mentation, this philosophical stance may be epitomized as follows: Science can only deal with what can be directly observed or what is subject to experimental verification. It is argued that failure to mark this precept historically led to science fraught with speculation, metaphysics, and even theology—witness the *élan vital* of Bergson, the entelechies of Driesch, and various theological teleologies that have perennially attempted to capture biology, from William Paley to Creationism. It is evident, the argument continues, that thoughts, feelings, concepts, desires, and intentions in animals are not the sorts of things that can be either perceived or explored experimentally. Thus such material is not a legitimate object of study. This position, implicit in some versions of positivism, found clear expression in Watson's formulation of behaviorism, and it exerted major influence even on positivist thinkers otherwise inimical to behaviorism, such as Konrad Lorenz and Nikolaas Tinbergen. (Thus, the 1948 volume *Instinctive Behavior* [Schiller, 1957], which chronicles the first encounter between behaviorists and ethologists, stresses the absolute concord between the two groups regarding the methodological need for eschewing talk of animal mentation. The two factions in fact agreed on little else.)

Clearly, anecdotal anthropomorphism, like all other forms of argumentation, is subject to abuse. For example, I cited the case of my wife's coworker who believed that her dog was capable of knowing when his birthday was and also of celebrating it. And indeed, there are many examples in the history of science of the excesses of anthropomorphism. For example, one of my cherished possessions is an early nineteenth-century, voluminous textbook of entomology, William Kirby and William Spence's *Introduction to Entomology*, written in elegant, belle-lettristic prose characteristic of an era far more literate than ours. In addition to describing the biology and behavior of a great variety of insect species, the chapters are sprinkled with bits of poetry descriptive of the insect's life form. Most bizarre perhaps is that the authors included moral teachings drawn from insect behavior, such as behavior depicted as exemplifying moral virtues such as thrift, timeliness, industriousness, and so on. While striking us as ridiculous, there is a certain charm in recalling an era where morality was juxtaposed with science in the same book. Many biological textbooks took the same form, and they are legitimately described as subject to the excesses of anthropomorphism. It is clearly a logical error to move from a condemnation of unbridled anthropomorphism to a righteous condemnation of *any* anthropomorphism.

Although, as we saw earlier, Charles Darwin saw anthropomorphism as an inevitable consequence of phylogenetic continuity of mentation, subsequent scientific ideology is considerably less astute and dismisses anthropomorphic reasoning as fallacious. In addition, Darwin collected voluminous amounts of anecdote illustrating animal mentation, which, as mentioned earlier, he later turned over to his secretary, George Romanes, who edited them into two volumes, *Animal Intelligence* and *Mental Evolution in Animals*. Romanes was scrupulous in his criteria for accepting anecdotes.

It is high time, Romanes tells us, that we realized that "the phenomena which constitute the subject-matter of comparative psychology, *even if we regard them merely as facts in nature,* have at least as great a claim to accurate classification as those phenomena of structure which constitute the subject-matter of comparative anatomy" (1882, vi, my emphasis). The phenomena of animal mentation are the subject matter of facts, facts that have been common coin throughout human history, facts that are in principle no more problematic than any other kind of facts. We can know them as we know the facts of human mentation. There are no epistemological chasms to be bridged or metaphysical barriers to be scaled. The data pertaining to animal thought has been accumulating since the dawn of humanity; the methodological problem is separating the wheat from the chaff, the same sort of problem that exists in all areas of human knowledge. Fresh impetus has been given to this enterprise, Romanes says, by the advent of evolutionary theory, which gives greater credibility to these facts of common experience, explains them theoretically, and points to the high probability of "genetic continuity" (1882, vi) between human and animal intelligence.

But if what ensues is simply the chronicling of anecdotes relating to animal mentation along the phylogenetic scale, how does Romanes differ from the anecdote-mongers whom he deplores? In the first place, he would doubtless say, in seriousness of purpose, and second, in attempting to place his anecdotes within the context of evolutionary theory. But third, and more to the point, he differs in the care with which he provides criteria for selection of the facts that he addresses. As he puts it, "Considering it desirable to cast as wide a net as possible, I have fished the seas of popular literature as well as the rivers of scientific writing" (1882, vii). Initially, Romanes had intended to countenance as facts only what had been reported by observers known to be competent; but soon he realized that this was too rigorous since the probability of the more intelligent individuals among animals happening to fall under the observation of the more intelligent individuals among men was extremely low. So instead, he looked to other, less restrictive principles:

First, never to accept an alleged fact without the authority of some name. Second, in the case of the name being unknown, and the alleged fact of sufficient importance to be entertained, carefully to consider whether, from the circumstances of the case as recorded, there was any considerable opportunity for malobservation; this principle generally demanded that the alleged fact, or action on the part of the animal should be of a particularly marked and unmistakable kind, looking to the end which the action is said to have accomplished. Third, to tabulate all important observations recorded by unknown observers, with the view of ascertaining whether they have ever been corroborated by similar or analogous observations made by other and independent observers. This principle I have found to be of great use in guiding my selection of instances, for where statements of fact which present nothing intrinsically improbable are found to be unconsciously confirmed by different observers, they have as good a right to be deemed trustworthy as statements which stand on the single authority of a known observer, and I have found the former to be at least as abundant as the latter. Moreover, by getting into the habit of always seeking for corroborative cases, I have frequently been able to substantiate the assertions of known observers by those of other observers as well or better known. (Romanes, 1882, viii–ix)

What is one to say of this method of Romanes, which to a working laboratory scientist would appear to be not much of a method at all? Where are the controlled experiments? Where are the hypotheses and tests thereof under controlled conditions? Where is the distinction between the observations of trained scientists and those of laymen? Where does Romanes come off making the dubious assumption that animals have mental traits at all, rather than merely being complex mechanisms? Hearsay! Anthropomorphism! Anecdote! Not to be taken seriously!

Some of these objections are dealt with rather presciently by Romanes himself. Responses to others can be extrapolated from the logic of his position. However, none are terribly difficult to deal with once we have deduced the consequences of his assumptions.

One of his assumptions is that, given evolutionary theory, it is hard to see how one can avoid postulating continuity of mental traits between humans and animals. Second, common sense across all ages and all cultures has seen and explained animal behavior in terms of mentalistic attributions. Romanes even has an argument to this effect, namely, that if we do not allow appropriate animal behavior to count as evidence of feeling and mentation, what right do we have to allow appropriate human behavior to serve as such evidence? The only consciousness to which we have direct access is our own, the minds of other people being as inaccessible as those of animals. If we can argue by analogy in the one case, we can surely do so, *mutatis mutandis,* in the other. But if we choose to be skeptical about animal minds because we have no direct access to them, we must extend our skepticism to the minds of other people as well. And not only to other people, but also to the external world, for, in the final analysis, all we have are our own perceptions, which do not certify the existence of an external world existing intersubjectively and out-side perception. But in that case, physical science is no more coherent con-ceptually than mental science; so the science of bodies is no more defensible than the science of minds. Insofar as ordinary common sense disregards this sort of skepticism as idle chatter, it must do so for mind as well as for body, animal or human. Romanes puts it this way:

> The only evidence we can have of objective mind is that which is furnished by objective activities; and as the subjective mind can never become assimilated with the objective so as to learn by di-rect feeling the mental processes which there accompany the ob-jective activities, it is clearly impossible to satisfy any one who may choose to doubt the validity of inference, that in any case other than his own mental processes ever do accompany objective activ-ities. Thus it is that philosophy can supply no demonstrative ref-utation of idealism, even of the most extravagant form. Common sense, however, universally feels that analogy is here a safer guide to truth than the sceptical demand for impossible evidence; so that if the objective existence of other organisms and their activities is granted—without which postulate comparative psychology, like all the other sciences, would be an unsubstantial dream—common sense will always and without question conclude that the activi-ties of organisms other than our own, when analogous to those activities of our own which we know to be accompanied by cer-tain mental states, are in them accompanied by analogous mental states. (1882, 6)

So much, then, for metaphysical objections. But what of the other ques-tions today's scientist might raise? Even if we grant that there are facts of animal mentation, surely they are best studied by controlled experiment, not by sifting through anecdotes.

Is Romanes's anecdote sifting scientifically invalid? Does it require that we weaken and suspend ordinary canons of proof? Quite the contrary. What Romanes recommends that we do with the vast hodgepodge of data rel-evant to animal mentation is to apply to it the same sort of reasoning we employ when we reconstruct historical events, write biographies, assess peo-ple's motives or their guilt and innocence in trials, defend ourselves against accusations, make judgments about people of whom we hear conflicting stories, and so on. In all these cases, what we do is to measure data against standard rules or canons of evidence and plausibility. Does the data violate known laws or established evidence, as in the case of the parrot reported by Locke to be able to hold conversations? Does the source of the data have a vested interest in telling a certain sort of tale? Does the source have an axe to grind? Does the data accord with other data from totally independent sources across time and space? In short, weighing data on animal mind is no different from weighing any other data, for example, data pertaining to the personality of some historical figure. Thus, we discredit Frau Hitler's assertion that Adolf was a nice boy who wouldn't hurt a fly. If you start out with the a priori assumption that no data is relevant vis-à-vis the existence and nature of animal thought, then such a method is absurd. But if you begin with the common-sense (and evolutionary) view that the existence of animal pain, suffering, guilt, planning, fear, remorse, loyalty, and other mental states is self-evident, and that what needs to be done is to sort out the scope and the limits of such states, then this method is not only plausible, but, given a sea of data, inevitable, in order to separate the wheat from the chaff.

But surely, says today's scientist, such random observations must lack the credibility of data garnered by trained observers under laboratory conditions. This, at least, we must surely concede, the scientist says. Not at all—no more so than we must concede that laboratory experiments provide a better guide to human motivation and human nature than ordinary experience. No controlled experiment will provide me with better evidence that one of my friends is a lecher or that people will cut corners to make money than does my ordinary experience. Laboratory experiments on animal thought (or behavior) tend to focus on abnormal animals under highly abnormal conditions. Though the laboratory rat and the cat are among the most highly studied subjects in twentieth-century psychological research, much of the data pertains to their behavior under the most unlikely conditions imaginable, as when they are being shocked, frightened, or presented with inescapable painful situations designed to create "learned helplessness"; or are crowded, blinded, confined, and so on. It is hardly surprising that little has come from all this that even begins to describe—let alone explain—"normal" cat behavior. In a key sense, all laboratory examination is by definition extraordinary and likely to miss an organism's normal behavior and is concomitantly unlikely to help us understand telos.

Given the formidable arsenal of arguments arrayed against talking anthropomorphically of mental states in animals and buttressing such claims anecdotally, one can understand scientists' reluctance to countenance such talk, and it is indeed the case that such talk virtually disappeared from scientific literature during most of the twentieth century. Nonetheless, the issue has once again been thrust forward into the scientific arena. Why has this occurred? There are a multiplicity of historical vectors that have militated in favor of softening the positivistic/behavioristic skepticism about animal consciousness. But one in particular is worth recounting here: ordinary common sense, as distinct from the common sense of science, has of course never caviled at mentalistic attribution to animals; indeed, as Hume points out, few things are more repugnant to ordinary common sense than skepticism about animals' mind. But until recently, ordinary common sense cared little about the implausibility of scientific common sense; if scientists wanted to believe that animals have no mind, so what: scientists believe many strange things.

A major clash between these two competing common senses has arisen only in the last few decades, for it is only in that period that ordinary common sense has begun to draw any significant *moral* implications from the presence of thought and feeling in animals. Although ordinary common sense certainly never doubted that animals could feel pain, fear, and so on, it drew no moral conclusions from this, largely because animal exploitation was invisible to daily life in virtue of the nature of animal use. Science, on the other hand, insulated itself from the moral implications of its own activity with animals not only by the denial of animal mentation but by the other mainstay of scientific ideology, the claim that science is value-free and thus can make no moral claims and take no moral positions, since moral judgments are unverifiable.

Of late, however, ordinary common sense has grown increasingly conscious of our moral obligations to animals and increasingly unwilling to let science go its own way. The reasons for this change in public attention to animal treatment and to science's agnostic attitude thereto are largely moral ones, growing out of profound changes in animal use that have arisen in the past sixty years. Prior to World War II, and indeed for virtually all of human history, the overwhelming use of animals in society was agricultural—animals were reared for food, fiber, locomotion, and power. The key to successful animal production was *husbandry,* an ancient term derived from the Old Norse word *husbondi,* meaning "bonded to the household."

Husbandry meant putting one's animals into the optimal environment in which they were biologically suited to thrive by virtue of natural and artificial selection, and further augmenting their natural ability to survive and flourish by provision of medical attention, provision of protection from predation and extremes of climate, provision of food and water during times of famine and drought, and so forth. *Husbandry was in essence keeping animals in ways acknowledging and respecting their teloi.* Indeed, so powerful is the husbandry imperative in human history that when the Psalmist seeks a metaphor to schematize God's relationship to humans, he draws upon the archetypical husbandry role, that of the shepherd:

> The Lord is my shepherd. I shall not want. He maketh me to lie down in green pastures, He leadeth me beside still waters, He restoreth my soul.

The husbandry imperative was thus an almost perfect amalgam of prudence and ethics. It was self-evident that "the wise man took care of his animals"—to fail to do so was to harm oneself as well as one's animals. Husbandry was assured by self-interest, and there was thus no need to place heavy ethical emphasis on proper care of animals. The one exception was the ancient prohibition against overt cruelty and outrageous neglect, designed to cover those rare sadists and psychopaths unmoved by self-interest.

Proper treatment of animals, then, for most of human history, was not heavily stressed in social ethics since it was buttressed by the strongest of motivations—self-interest. Husbandry-based animal agriculture—the overwhelming majority of animal use in society—was about putting square pegs in square holes, round pegs in round holes, and generating as little friction as possible while doing so. Animal agriculture—historically virtually *all* animal use—was thus a fair contract between humans and animals, with both sides benefiting from the ancient contract represented by domestication.

This ancient and fair compact changed dramatically in the mid-twentieth century with the rise of high-technology agriculture. With the advent of what I call "technological sanders"—antibiotics, vaccines, hormones, and so forth—one was no longer constrained in one's agriculture by the animals' biological natures. One could now force square pegs into round holes, round pegs into square holes, with the attendant animal suffering irrelevant to profit. The connection between animal welfare and animal productivity was severed. Similarly, with the rise of massive amounts of research and toxicity testing on animals beginning at approximately the same time, animal use could benefit us while harming them in unprecedented ways— inflicting disease, wounds, burns, fear, pain, and so forth on animals so we could study them, with no compensatory benefits to the animal subjects. For the first time in history, the welfare of animals used by humans became a moral issue. By the late 1970s, Europeans and North Americans were demanding that animal use be modified in research and agriculture so that suffering would be mitigated and animal well-being would be assured.

In this way scientific ideology, agnostic about animal consciousness, clashed with ever-increasing social concern about animal treatment. This new social tendency to concern itself about animal welfare forces upon science what I have described as "the reappropriation of ordinary common sense" about animal thought and feeling. Thus, for example, in the face of federal law that mandates control of pain and suffering in laboratory animals, it is obviously inappropriate for scientists to express total skepticism about our ability to know what animals think and feel. Thus scientific ideology is now threatened and must bend to accommodate ordinary common sense.

Take, for example, the symposium on animal pain and suffering convened by the American Veterinary Medical Association (AVMA) in 1987 and its attendant panel report (American Veterinary Medical Association, 1987). The report acknowledges that animals do feel pain, pointing out that pain research that is extrapolated to humans is after all done on animals, which presumes that they feel pain. (Traditional scientific common sense had explained pain research as research into pain mechanisms and behavior, and had ignored any talk of the subjective experiential dimension.) Indeed, the report continues quite reasonably, all animal research that is used to model human beings is based in a tacit assumption of anthropomorphism; and if one can in principle extrapolate from animals to humans, why not the reverse as well?

But a hard-line proponent of the common sense of science would very likely remain unmoved by our discussion, and he or she might respond as follows: Granted that political pressure forces upon us the need to behave as if animal consciousness is scientifically knowable and assumable. But in fact, it is not, for the reasons detailed above.

A bitterly amusing example of such scientific recalcitrance was related to me by Robert Rissler, the US Department of Agriculture official charged with interpreting the 1985 federal laws aimed at furthering the welfare of laboratory animals. In addition to mandating the control of pain and suffering through proper anesthesia, analgesia, sedation, and euthanasia, the laws required that nonhuman primates used in research be provided with environments that "enhance their psychological well-being." Rissler, a veterinarian, knew little about primates and even less about their psychological well-being, having himself been trained under agnosticism about animal mentation. Nonetheless, he was charged with writing regulations giving operational meaning to the psychological well-being of primates. Somewhat naively, he approached the American Psychological Association's Primatology Division, seeking their counsel on defining this obscure notion. "Don't worry," they assured him. "There is no such thing." "Oh, but there will be after January 1, 1987 [the date the laws took effect], whether you help me or not!" replied Rissler tellingly.

Science is, of course, our vehicle for knowing about the world. If science denies our ability to access animal mentation through anecdotal data and anthropomorphic locutions, it removes itself from answering or helping to answer the key ethical questions about animal well-being that have emerged in society. To address questions of animal treatment, welfare, acceptable environments, pain and suffering, and so forth, we must be able to make meaningful claims about what animals experience and feel.

To do this, we must in turn be allowed to use anthropomorphic locutions based in our ordinary empathetic experience of animals' lives. My animal agriculture students, when taught animal behavior by a mechanistic teacher who refused to use mentalistic locutions about animals, reported ignoring the professor's teachings when they went home to their ranches. "If I can't say that the bull is pissed off today," said one such student, "I won't live real long." Our ability to work with animals, anticipate their behaviors, and meet their needs rests foursquare on such locutions. Scientific common sense's agnosticism about such locutions therefore in essence removes questions of animal welfare from the realm of legitimate empirical investigation.

It is necessary to point out that the skepticism discussed above, if systematically adhered to in science, would render doing science impossible. For in actual fact, presuppositional to scientific activity are certain assumptions that flagrantly violate the claim that everything in science must be observable or subject to direct experimental confirmation. Consider the following: Science assumes that there is a real, public, intersubjectively accessible world out there, existing independently of my perceptions, and accessible to other humans and to other scientists in particular. It also assumes that other scientists perceive the public world and think more or less as I do, and that one can distinguish veridical and falsidical scientific reports about experiences of that world. It also assumes that there really is a past, even though we cannot experience it directly, and that it is not the case that the universe was created three seconds ago, fossils and all, and us with all our memories. The key point is that none of those beliefs can even in principle be confirmed by observation or directly tested by experiment. Yet few scientists are disposed to reject them, even though they conflict with what is entailed by the assumptions of scientific common sense. If they did reject them, they could not do science. (What would solipsistic science be like? Why publish?) Thus, the hard-line skepticism discussed above must be tempered, else it destroys science altogether.

The obvious response is that it is wildly *implausible* to embrace solipsism, to deny other minds or an external world, or to treat the history of the world as being three seconds in length. And to this I fully agree. And, in my view, it is *equally* implausible to deny mentation to animals. Philosophically, as soon as one has given up a hard-line verificationism that admits only direct observables into science, and one has admitted that certain nonverifiable beliefs are admissible on the grounds of plausibility (e.g., belief in an external world independent of observers and commonly accessible), one has replaced a rigid logical criterion for scientific admissibility with a pragmatic one, in which one needs to *argue for* exclusion of certain notions from science rather than simply apply a mechanical test. And, of course, this is what has in fact occurred in the history of science—science has talked of all sorts of entities and processes that are not directly verifiable or directly tied to experiment, from gravitation to black holes. Indeed, contemporary physics, traditionally cited as the hardest of hard science, has positively proliferated notions that violate the common sense of science. Such theoretical notions are accepted, of course, because they help us understand reality far better than we do without them.

Talk of mind in animals has a similar justification. We have already mentioned psychologist David Hebb's point that we could not interpret animal behavior in ordinary life without imputing such notions as pain, fear, anger, and affection to animals, all of which have a mentalistic component in addition to a behavioral one. For saying that "when a dog is in pain that means only that the dog is exhibiting a certain range of behaviors or responses" does not explain its cringing or loss of appetite unless we also assume that it is *feeling* something—"hurt"—which is functionally equivalent to what we feel when we hurt. This assumption is, in fact, as the AVMA panel report on recognition and alleviation of animal pain said, presuppositional to doing pain research and analgesia screening in animals and extrapolating the results to people. What we are interested in is a *feeling* common to both, not merely similarity in plumbing and groaning.

I have thus far attempted to establish that the traditional scientific skepticism about animal mind is wrong-headed. Furthermore, using the example of pain, I have argued that, in at least some cases, scientific attribution of mentation is inevitable and based in anthropomorphism as presuppositional to its intelligibility.

It is now relevant to reintroduce the notion of anecdote as a source of information about animal mentation and to assess its relevance to science. An excellent place to begin, for it retains the simple case of pain we have been using, is a famous article by David Morton and P. H. M. Griffiths that appeared in the *Veterinary Record* in 1985. This article was one of the first papers addressing the recognition and alleviation of pain in animals. It is noteworthy that while the authors do provide criteria for assessing pain and its degree in animals, they stress that the best sources of information about animal pain are farmers, ranchers, animal caretakers, trainers—in short, those whose lives are spent in the company of animals and who make their living through animals. Given the plausibility criterion discussed earlier, the advisability of seeking information on pain from animal caretakers is patent. Whereas scientists could get on perfectly well in highly artificial laboratory situations professing agnosticism about animal pain and other mentation, those who live with and depend on animals could not. If you fall into the latter class and do not recognize pain, fear, anger, and so on in your animals, you will lose your livelihood, be highly vulnerable to injury, and be unable to control or train your charges, among many other negative outcomes.

Thus, given that science specifically disavowed the reality of animal thought and made no attempt to study it, it is perfectly proper to look to those who have been *compelled* to understand animal thought for millennia. To be sure, such information will be "anecdotal," that is, not obtained in laboratory experiments and not analyzed, but that does not mean it is illegitimate.

Thus we have seen that, in the simple case of pain, the common sense of science is wrong, and that one can talk of what animals experience; that one must use a measure of anthropomorphism, even as we use our own individual experiences as a guide to understanding those of other humans; and that one must depend, at least currently, on anecdotal information. Indeed, an even more striking argument could be made regarding the concept of suffering, which does not appear in the scientific literature even with regard to humans, let alone animals.

One can also buttress these arguments with others. Similar physiological mechanisms for pain in humans and animals, similar behavioral responses, similar neurochemistry, and the plausibility of phylogenetic continuity all militate in favor of attributing felt pain to animals, as does the fact that humans who do not feel pain, for congenital or acquired reasons, do not fare well.

One could respond to the argument we have developed thus far in this way: As long as one focuses on simple, fundamental, primitive mental experiences like pain, one's argument is unexceptionable. But as soon as one leaves sensation and begins to talk of higher mental processes in animals, one cannot accept anecdotal anthropomorphic evidence. Ordinary common sense and its discourse are far too disposed to exaggerate animal intelligence, planning, reasoning, and emotional complexity and to jump to unwarranted conclusions by seeing animals as furry humans. Indeed, it was precisely romantic, unbounded anthropomorphism and exaggerated anecdotes' abounding in the nineteenth century that in part led to the behavioristic/ positivistic reaction against animal thought.

How does one reply to such an objection? In the first place, one might argue (as did F. J. J. Buytendijk in *Pain: Its Modes and Functions*) that the ability to feel and respond appropriately to pain bespeaks mental sophistication beyond mere sensation. Pain in and of itself would be of little value if it were not coupled with some ability to choose among alternative strategies of response, for example, fight or flight, hide, evade, and so on. Thus, the evolutionary utility of pain consists in the organism's ability to respond to a noxious stimulus not only with motivation to alleviate it but with strategies to deal with it as well. It is in fact this insight that led pain physiologists Ralph Kitchell and Michael Guinan to conjecture that animals may well suffer more from pain than humans do. Since animals lack the cognitive abilities possessed by humans to understand the sources of pain and to formulate strategies for its relief or for its mitigation, Kitchell and Guinan (1990) suggest that the motivational dimension of pain, that is, the hurting and the correlative drive to escape it, may well be *more profound in animals than it is in humans, and thus their experience of pain may be, in balance, worse than ours.* As I have pointed out elsewhere in this regard, it has often been argued that in lacking the tools for transcending the here and now provided by linguistic capacity and concepts, animals correlatively lack the suffering occasioned in us by anticipation of pain and other noxious experiences. However, if this is the case, they also lack anticipation of the end of pain and thus have no *hope.* In a terrible way, they are their pain; there is no light at the end of the tunnel for them.

Be that as it may, I should rather respond by affirming that the argument and strategy we have constructed for using anecdotal and anthropomorphic information to identify pain is in principle no different from using the same approach to understand higher (or other) mental processes. The relevant distinction is not pain (or sensation) versus thought (or higher mental processes)—it is rather good versus bad anthropomorphism, reasonable versus unreasonable anecdote.

Once again, the key notion for our analysis is *plausibility,* the same sort of measure we use when we attribute thoughts, plans, feelings, and motives to other humans, be it in daily life or when serving on a jury. Let us recall that we do not experience other people's mental states and that language can be used to conceal and deceive. How, then, do we judge other humans' mental states? What we do is use a combination of weighing of evidence and what we might call "me-thropomorphism"—extrapolations from our own mental lives to others'. In other words, for example, if a friend of mine who has evidenced a normal propensity for jealousy suddenly finds his wife, whom he has loved deeply, running around flagrantly with another man, and tells me he bears her no ill will, I am skeptical. I can in principle be convinced that he feels no jealousy, but this would require very extensive observation and interaction with him to trump my plausible interpretation. On the other hand, if he tells me that he *is* jealous and angry, or behaves that way, it is certainly reasonable to assume that he is, for the motivation for that response is what I know of myself and others. Although we would be hard pressed to articulate them, we all have canons for judging the plausibility of anecdotes about other people and of explanations of their behavior. The female student who tells me that a male professor obviously has a crush on her and cites as evidence the fact that he ran into her twice in one week at the grocery store and said "Hi" may reasonably not be taken seriously. While I can understand how a person could take such events to mean something, and while the student's conclusion could in fact be true, in my consideration of the evidence given in this anecdote, I would likely decide that it is not. In fact, the vast majority of our knowledge of human behavior does not come from scientific research but from our evaluation of life experience and what would be dismissed by the common sense of science as "anecdotal."

My claim, like that of Darwin's secretary, George Romanes, is that anecdote is, in principle, just as plausible a source of knowledge about animal behavior as it is about human behavior, provided it is tested by common sense, background knowledge, and standard canons of evidence. Thus, when a child says that there is someone on the street giving away money to everybody who wants it, we suspect either misunderstanding or swindle since, by and large, people do not do such things. By the same token, when someone interprets his dog's restlessness as evidence that the dog knows it is his birthday, we can dismiss that since we have no reason to believe that the dog has, or even can have, a concept of birthday. On the other hand, if the person telling the anecdote explains the dog's excitement by saying that he has learned that when his owners cook and clean all day and frequently look out of the window, guests are coming, that is consonant with what we know of dogs' abilities.

More difficult cases occur when the anthropomorphic anecdote concerns a species of animal with which we do not enjoy the familiarity we do with dogs, though here the problem is in principle no different from when we deal with people who come from cultures significantly different from ours. When they belch loudly after a meal, we may label them as rude people out of ignorance of their culture wherein such an act is a polite compliment; we can make the same mistake looking at unfamiliar animals. Witness the child or urban adult who reports equine sex play as fighting. Thus it would seem to me that once one has in principle allowed the possibility of anthropomorphic, anecdotal information about animal mentation, one must proceed to distinguish between plausible and implausible anecdotes, the latter of which may nonetheless turn out to be true, though we are right to be skeptical, and likewise between plausible and implausible anthropomorphic attributions.

The fact that many people tell outrageous anecdotes or interpret anecdotes in highly fanciful or unlikely ways, and even publish such nonsense, should no more blind us to the plethora of plausible anecdotes, and reasonable interpretations thereof, forthcoming from people with significant experience of the animals in question, than should the presence of outlandish stories about or outrageous interpretations of human behavior cause us to doubt all accounts of human behavior.

Anecdotes and their interpretation may obviously be judged by many of the sorts of principles Romanes relates in his classic introduction to *Animal Intelligence*. Does the anecdote cohere with other knowledge we have of animals of that sort? Have similar accounts been given by other disinterested observers at other times and in other places? Does the interpretation of the anecdote rely upon problematic theoretical notions? (Cf. imputing a grasp of "birthday" to the dog.) How well does the data license the interpretation? Does the person relating the anecdote have a vested interest in either the tale or its interpretation? (Penny Patterson's stories of Koko the tame gorilla in her fund-raising letters naturally excite some skepticism.) What do we know of the teller of the anecdote—is it Konrad Lorenz or Baron Munchausen? One can—and we do—set up plausible rules for judging anecdotal data, be it about humans or about animals. The alternative is to create total, nihilistic skepticism about the commonsense experience that has given us most of our social knowledge of the behavior of people and animals.

One fascinating point that has escaped notice is that anecdotes are logically no worse off than reports of scientific experiments and their interpretation; in some ways the latter may be more suspect. As we know from the rash of reports of data falsification, fraud, and dishonesty in scientific publications, scientists are as human as anyone else. Given a "publish or perish" system for science, scientists feel the pressure to produce or else they must effectively give up their careers. If this is so, then researchers have a strong vested interest in obtaining results, which in turn should excite our natural suspicion of their results. It is true that scientific reports are replicable in principle, but there is little money for such replication as long as a result coheres with data in the field. Anecdotes are, of course, also replicable in principle, either by experiment or by observation. In the final analysis, *any report of an experiment is by definition an anecdote, not a confirmed hypothesis.* The following questions should be mulled over by anyone interested in these issues: Is it more unreasonable to trust the account of a disinterested lay observer or that of a scientist who must get results to survive? Is the multiplicity of theoretical biases that scientists carry in virtue of their training any less, or any more, or equally pernicious to their observational capacity than the theoretical biases built into a nonscientifically trained but intelligent observer?

Let us conclude with, appropriately enough, an anecdote that is very interesting to laypeople and to students of animal behavior. The story was in fact reported, in detail, on Denver television, accompanied by videotaped pictures of the events described. In the story, an African elephant at the Denver Zoo had gone down and refused to get up, a condition known to lead to fatality if not corrected. All efforts to get the elephant to stand up—including bringing in a hired crane—had failed. By chance, the Asian elephants were herded past the afflicted elephant. The Asian elephants broke ranks, approached the fallen elephant, and nudged and poked him until he stood up. They then supported him until he stood on his own.

Thus far, we have an anecdotal narrative, with little or no theoretical bias obtruding and no interpretation offered. As data relevant to the study of elephant behavior, the story is surely relevant. Although the TV station has a vested interest in dramatic stories, it filmed the events and its account was buttressed by other observers, so falsification of the events is unlikely. The commonsense interpretation of the data offered by the station, and by the average observer, was that the elephants were altruistically *helping* another elephant, albeit a different species. Such an interpretation is more problematic than the simple reportage of the events, since "help" is ambiguous and speculative. The events are certainly open to other interpretations. When, however, one juxtaposes that story with the many other stories of elephants' showing helpful behavior to other elephants, together with the extensive data we have on the problem-solving ability and the social nature of elephants, the interpretation gains in plausibility.

To preclude data on (and interpretations of) animal behavior a priori simply because the data was not garnered in laboratories (which constitute in any case highly unnatural conditions for animals) or was not observed by "accredited scientists" is against the spirit of what science should be. To be sure, common sense is "theory-laden" with often problematic categories and interpretations, but so too is science. It is at least as hard to see how intelligent, educated scientists bought whole-hog into behaviorism for most of the twentieth century as it is to see how ordinary people can buy into astrology today. As twentieth-century philosopher Paul Feyerabend suggests, science should be democratic in its admission of data sources but stricter in the theories or explanations it graduates.

Mentalistic attribution to animals provides a very plausible theoretical structure for explaining and predicting animal behavior. Anthropomorphism, if tested against reasonable canons of evidence, is another plausible—and indeed inevitable—theoretical approach to assessing animal behavior. And finally, since there are and always have been far more ordinary people out observing animals than there are scientists engaged in the same activity, it would be a pity to rule out anecdote, critically assessed, as a potentially valuable source of information and interpretation of animal behavior. In fact, ever-increasing social-ethical concern about animal treatment essentially requires information about what matters to animals as the raw material for formulating social policy. Since the common sense of science has morally castrated the language it uses to describe animal behavior, eschewing, for example, morally laden descriptions of animals as expressing pain in favor of "neutral" locutions like "vocalizing," the gap must be filled by the language of ordinary common sense, replete as it is with morally relevant locutions about animal experience.

In sum, we may now affirm that, for ordinary common sense, emerging animal ethics dovetails with, and indeed follows from, our social-consensus ethic, by way of telos. We have also seen that judging telos, and what matters to animals, is not problematic to ordinary common sense and that one can mount strong arguments against the skepticism emerging from scientific ideology.

Animal Telos and Animal Welfare

IT HAS BEEN a long and circuitous excursus to bring us to the point that most ordinary people begin with, namely, that animals have biological and psychological natures that we can know. It is important, though, to have done this in order to quiet ideologically based skepticism about the knowability of telos that might arise regarding the basis for animal ethics that we have developed in our discussion. Such skepticism inevitably arises from scientific ideology, or the common sense of science, which does not credit the world of ordinary experience as being the real world. But both the Aristotelian worldview and common sense do see the world we experience as real. Whereas, as we saw, a Descartes goes out of his way to discredit ordinary experience to prepare the way for an alternative view of reality, neither ordinary common sense nor Aristotle entertains systematic doubt about what we experience. For both, the validity of what we experience is the default—we may experience hallucinations, delusions, illusions, and deceptions, *but it is built into our nature to be able to detect and correct them.* Thus, neither Aristotle nor ordinary common sense begins with a *problem of knowledge,* wherein it is methodologically necessary to prove that humans can know the world. That is self-evident. Consider the resounding opening sentence of Aristotle's *Metaphysics*: "All men by nature desire to know." Implicit therein is the assumption that we *can* know.

For Aristotle, knowledge of the world derives from careful observations, critically assessed. If, for example, I wish to know what sort of maternal behavior is natural to pigs, I simply observe numerous instances of such behavior under natural conditions. There is no set number of observations required to create such knowledge. I may need to observe ten instances of

97

such behavior before my mind grasps the universal implicit in the particular; you may be able to apprehend the truth after six relevant observations. That witnesses a difference in the acuity of what Aristotle calls our "active intellects," or our ability to abstract, but both you and I eventually grasp a universal truth about swine maternal behavior. If I generalize too quickly, I am corrected by further observation. Grasping truths about the world is, to misquote Stokely Carmichael, "as natural as apple pie." Knowing is not a problem; it is a biologically determined fact for human beings. Unlike post-Cartesian science, there is no reduction of something to something else in the way that modern science aims to reduce phenomena to what can be expressed in mathematical equations. We know something (a fact) when we see that it is an instance of a universal law, at which point it becomes what Aristotle calls a "reasoned fact."

My thesis, then, is that animals have needs and desires flowing from their teloi that, when thwarted, frustrated, or simply unmet, result in negative feelings that are the experience of poor welfare. The scientific literature did not historically credit these feelings as real. For many years, it spoke instead of the "stress response" in an effort to remain scientifically solid by not invoking mental states but instead talking in physiological terms about secretion of cortisol under stressful conditions. This was the dominant approach despite the fact that by 1971, in the heyday of a physicalistic approach to stress, John Mason pointed out that over a hundred papers had demonstrated the strong effect that *purely psychological stimuli have on the physiological stress response.* Even more destructive to the coherence of a purely physiological approach to stress were the papers of Mason and Jay Weiss that appeared in the early 1970s. Mason (1971) showed that postulating mental states is necessary to explain the results of experiments on animals. He performed experiments in which he put animals under different conditions in which the physical stressors to which the animals were subjected were the same but the emotional, psychological, or cognitive states of the animals or the attitudes of the animals toward the stressor were different, and these states led to radically different physiological signs of stress—that is, radically different levels of secretion of the urinary steroid 17-OHCS. This shows that an animal does not respond to a noxious stimulus merely mechanically and uniformly, but that what makes a stimulus noxious or stressful to an animal is the animal's conscious "reading" of what is happening to it, its cognitive state, its emotional attitude, *the understanding of all of which requires the imputation of mentation to the animal.*

The work of Jay Weiss (1972) showed that the degree to which an invasive stressor affects an animal, physically and psychologically, has a strong relationship to the animal's mental state, emotional and cognitive. Weiss demonstrated that rats who could predict an electric shock because it was consistently preceded by an audible tone developed virtually no gastric ulcers, whereas other rats, receiving the same shock but with the tone sounded randomly with respect to the shock, developed a great deal of ulceration. Similarly, rats who could control the shock, and thus cope with the noxious stimulus, developed far fewer ulcers. These experiments bespeak the explanatory power of attributing mentation to animals; indeed, they cannot be explained without such postulation. This is consonant with the barbarous work done in the United States on learned helplessness, which shows that animals receiving inescapable shocks inevitably develop a state of pathological withdrawal and inaction. This work is morally defended by researchers on the grounds that it models human depression—another eloquent attestation to the theoretical relevance of imputing mental states to animals. (This defense, of course, ignores the obvious fact that if these states are relevantly analogous to noxious human states, one needs to ask what right we have to inflict them on animals when we would never do so to humans.)

Thus, what conditions animals are kept in and what manipulations we subject them to ramify in positive and negative mental states, which in turn determine the effect of these conditions and manipulations on animals' welfare. Such a claim was anathema for scientists in most of the twentieth century. We have just seen that the research of Mason and Weiss buttresses this claim. The question then arises as to why such research did not overturn ideological denial of consciousness. The answer is that, in general, as Mason himself pointed out, research done in the fields of psychology and psychiatry (in which he worked) is totally invisible to the standard loci of stress research—physiological laboratories and animal science departments. The invisibility of some areas of research to others to which it is highly relevant is an important, but generally unnoticed, impediment to scientific progress.

Let us apply our analysis to a simple case of animal welfare: One of the biggest current animal-welfare controversies concerns housing breeding sows in gestation crates, small cages two feet wide by three feet high by seven feet long. Today, the vast majority of swine produced in the United States (well over 90 percent) are born to pregnant sows who carry out their pregnancies or gestation periods (three months, three weeks, and three days) in these cramped conditions. Over the last decade, the Humane Society of the United States, the wealthiest and most powerful animal-welfare organization in the country, has run a series of referenda in twelve states aimed at eliminating these sow stalls, as well as battery cages for egg-laying chickens and solitary crates for veal calves. The referenda have passed in every state by a comfortable margin. (Recently, I was able to convince Smithfield Farms, the world's largest pork producer—with eighty million sows in multiple countries—to eliminate sow stalls for social-ethical reasons.)

I will never forget the first time I went to a sow confinement barn. Each animal lived in solitary confinement, unable to turn around, stand up straight, or, in the case of large sows, even to lie down fully stretched out. The animals were compulsively chewing the bars of their cages and salivating, with a frenzied look in their eyes. A few years later, in Ontario, Canada, I witnessed an even more dramatic scene. A swine farmer had a gestation-crate barn with which he enjoyed good financial success. His disillusionment with crates gradually grew until he felt compelled to build a group-housing barn, as he felt guilty about keeping these animals in full and severe confinement. The new barn was separated from the old barn by a ten-foot walkway. When I entered the old barn, in an effort to escape me the sows crashed backwards in their cages as much as they could and stood rolling their eyes and chewing the bars, expressing fear in the sounds they made. I then crossed over the walkway into the new barn, which was built without cages but rather one large open space separated by baffles into rooms. I walked up to the fence surrounding the enclosure and delightedly watched the sows walk over to me as I reached over to pet them, as they all the while made clearly happy noises. The farmer walked over to me with a look of intense embarrassment on his face. I had been lecturing prior to my arrival at the farm and was thus wearing a tie and jacket. "Sir," he said, "do you value that tie?" "It is a nice tie," I replied. "Why do you ask?" He pointed at my upper chest, where two sows were happily chewing their way up the two parts of my tie, thoroughly soaking it with pig saliva. So delighted was I to see the animals behaving normally and without fear that I let them continue to chew the tie! Both this farmer and another one in the same area had converted the sow barns to group housing out of a sense of moral obligation to the animals. Both had excellent stockmanship and husbandry skills, and both were able to make as much money with group housing as they had been able to do with confinement.

On another occasion, I was visiting a Niman Ranch facility, where the animals are not confined at all except by an electric wire that keeps the piglets off the road. One of the sows, who had recently had a litter of piglets in a straw nest housed within a large hoop barn, approached a group of human visitors. Sows in confinement will threaten and even attack humans who approach their piglets. In this case, the sow accompanied us to see the piglets, all the while making happy barking noises. (Pigs have a large repertoire of sounds they use for communication.) The farmer told us that she was proud of her litter, and showing off. After acting suitably impressed, we turned to leave. The sow planted herself in front of the farmer and began to nudge him until the two of them reached the automatic waterer, where she stood still. The farmer checked the waterer and found that it was clogged. There was no question as to what the sow had communicated and that she knew that he was in charge of the waterer.

The intelligence of pigs is legendary among people who work with them. At Colorado State University, we had a technician who raised Yucatan pigs (small, docile pigs who once lived in the Yucatan Peninsula in Mexico in people's houses) for a research laboratory. Impressed with their intelligence, she took one to dog obedience school, where he proceeded to win every possible award. She ended up taking the pig to the *Tonight Show* with Johnny Carson, where the animal caused a major sensation. Shortly thereafter, she was forbidden to publicize these pigs by the laboratory-animal company that paid her, which essentially chided her for making these pigs too lovable.

In any case, before discussing the telos of the sow in relation to gestation crates, it is worth making some general remarks about the majority of animals we as humans interact with. It does not take a Konrad Lorenz to figure out that any animal born with bones, muscles, tendons, ligaments—the whole machinery of motion—is built to move. To argue for this concept is to invite astonishment at the need to do so, and even ridicule, from ordinary people.

The fact is, however, that modern industrialized agriculture has developed by keeping animals in severe confinement for economic reasons. Thus, anyone attempting to change some of these contemporary high-confinement systems must do the animal-behavioral equivalent of reinventing the wheel. Historically, the pig was the first farm mammal to be subjected to extremely intensive, or highly confined and industrialized, housing and management, a trend that has greatly accelerated. Over 90 percent of pigs are now raised in some kind of confinement. At the same time, swine are almost universally considered the most intelligent of farm animals, possessed of a good deal of curiosity and learning ability, as well as a complex behavioral repertoire. The complexity of pig behavior raises a host of issues relevant to rearing these animals under austere confinement conditions. Such conditions lead to a significant range of behavioral anomalies in confined pigs, referred to in the industry as "vices." As I have pointed out, such a locution is misleading and downright inaccurate, for it suggests that the pigs are somehow to blame for the aberrant behavior they display under confinement conditions. The animals' behavioral anomalies result from an attempt to cope with conditions that frustrate their nature, or their telos. Ronald Kilgour and Clive Dalton note in their *Livestock Behaviour: A Practical Guide,* "Pigs are easily bored and housing and management should be planned to provide for their inquisitive nature. This will prevent most vices, which are the result of boredom" (1984, 187).

Thus in the swine industry one encounters a host of welfare problems that are the direct result of the industrialization of agriculture and are based on thwarting the animals' behavioral and psychological needs and their nature. In general, the behavior of domestic swine is not far removed from that of their wild counterparts. Many cases are known in the United States of groups of domestic pigs that became feral populations and displayed the entire behavioral repertoire shown by wild swine that had never been domesticated. Moreover, controlled studies show that pigs born and reared in confinement, descended from generations of other confinement-born and -reared animals, will display "natural" pig behavior when placed in extensive conditions—pasture-based agriculture where animals graze freely and are not closely confined—for example, they will head directly for a mud hole and wallow. On the basis of such work, and other research into swine preference, one can get a sense of the full range of pig behavior and begin to understand the most serious areas of deprivation in confinement.

A summary of "natural" swine behavior and preferences, or a description of the animals' telos, can serve as a guide to identifying problematic areas in the confinement-agricultural rearing of swine. The most exhaustive study to determine this behavior was performed at Edinburgh University in the early 1980s by D. G. M. Wood-Gush and Alex Stolba, who placed domestic pigs in a "pig park," essentially a large enclosure replicating conditions under which wild swine live (Wood-Gush and Stolba, 1981): the enclosure contained a pine copse, gorse bushes, a stream, and a swampy wallow. Small populations of pigs, consisting of a boar, four adult females, a sub-adult male and female, and young up to about thirteen weeks of age, were studied over three years. The researchers observed not only the behavior patterns of the animals but also how the pigs used the environment in carrying out their behavior.

It was found that the pigs built a series of communal nests in a cooperative way. (This flies directly in the face of the confinement industry's claim that sows will "fight to the death" if housed in groups.) These nests displayed certain common features, including walls to protect the animals against prevailing winds and a wide view that allowed the pigs to see what was approaching. The nests were far from the feeding sites. Before retiring to the nest each night, the animals brought additional nesting material for the walls and re-arranged the nest.

On arising in the morning, the animals walked at least seven yards before urinating and defecating. Defecation occurred on paths so that excreta ran between bushes. (In outrageous violation of this natural tendency, pigs in sow stalls must urinate and defecate in the small space in which they eat and sleep.) Pigs learned to mark trees in imitative fashion. The pigs formed complex social bonds between certain animals, and new animals introduced to the area took a long time to be assimilated. Some formed special relationships—for example, a pair of sows would join together for several days after farrowing and forage and sleep together, contrary to the industry's self-serving claim that when sows are put together, they fight. There may indeed be skirmishes for establishment of dominance, but the hierarchies are quickly worked out. Members of a litter of the same sex tended to stay together and to pay attention to one another's exploratory behavior. Young males attended to the behavior of older males. Juveniles of both sexes exhibited manipulative play. In autumn, the pigs devoted more than half of the day to rooting.

Pregnant sows chose nest sites several hours before giving birth a significant distance from the communal nest (almost four miles in one case). The sows built nests, sometimes even with log walls. The sows did not allow other pigs to intrude for several days but sometimes eventually allowed another sow with a litter, one with whom she had previously established a bond, to share the nest, though no cross-suckling was ever noted. Piglets began exploring the environment at about five days of age and weaned themselves at somewhere between twelve and fifteen weeks. Sows came into estrus and conceived again while lactating.

One of Wood-Gush and Stolba's comments is telling: "Generally the behavior of pigs, born and reared in an intensive system, once they had the appropriate environment, resembled that of the European wild boar" (197). In other words, there is good reason to believe that domestic swine are not far removed from their nondomestic counterparts. Thus, comparison of behavioral possibilities in confinement with those in the rich, open environment that pigs have evolved to cope with seems a reasonable way to at least begin to assess the welfare adequacy of confinement systems. If confined environments generate behavioral disorders in the animals, this represents additional reason to believe that there are problems with these environments. Using the results of this study, we are in a good position to measure current systems against all aspects of the swine telos, physical and behavioral.

Those who defend sow stalls argue that they allow people who may not be "pig smart," as one expert puts it, to work in a facility where the system compensates for lack of stockmanship. On the other hand, management makes the difference between a viable confinement system and a total mess.

Unquestionably, such systems have produced large volumes of pork at reasonable prices. And as confinement managers argue, there are certainly benefits to the animals. In parts of the country where there are extremes of temperature—for example, Colorado and the Midwest, where temperatures may range from minus 20 degrees to 100 degrees Fahrenheit in the course of the year, and where rain, snow, and wind can make life miserable—an environmentally controlled habitat set for the animals' comfort zone can be a boon. In addition, individual, as opposed to group, housing of sows cuts down on fighting and biting and thus on wounds and competition for food.

Nevertheless, the costs to the animal in terms of its telos are considerable, especially in the face of what we know of swine behavior under the extensive conditions in which pigs have evolved. Most obvious to the agriculturally naive observer, perhaps, is the lack of exercise. As stated earlier, one need not be an animal behaviorist to realize that all animals who have evolved with bones and muscles need the opportunity to use them. As seen in our discussion of swine behavior in the Wood-Gush and Stolba study, pigs under extensive conditions spend a good deal of time moving about. If a system does not allow such an animal even the room to turn around, it is reasonable to view it as thwarting some very fundamental needs or tendencies, needs that have both a physical and a cognitive component, thus leading to negative welfare. Animals who like to move and are built to move are surely affected negatively if they cannot do so. Let us also recall that swine have been domesticated for ten thousand years: over 99 percent of that time was spent under pastoral conditions, and the animals fared well without sow stalls.

Closely connected with the inability to move is the element of monotony, lack of stimulation, or, as the ethologists Kilgour and Dalton (1984) forthrightly put it, boredom. Given the complexity of behavior and intelligence natural to the sow, the absence of possibilities in the gestation stall, and the emergence of stereotypies in the confined sow, it defies good sense to suppose that the animal is *not* bored. And I frankly do not cavil at the use of a word like *boredom,* which the scientific ideology has shunned as anthropomorphic. First, as an elegant article by the ethologist Francoise Wemelsfelder has detailed at great length, one can provide precise scientific meaning to terms such as *boredom* applied to animals (Wemelsfelder, 1985). Second, and most important, the common sense that informs social ethics will never doubt that sows are capable of being bored and will judge the system accordingly.

According to the farm-animal behaviorist Joseph Stookey,

> Such issues as sow stalls are destined to be resolved by moving in the direction towards systems which the public finds acceptable. Obviously the public is concerned about restricting the freedom of movement of animals and the public would prefer that such systems (such as tethers and sow stalls) be abolished. This is a response that has nothing to do with "welfare" from the pig's perspective (though they would argue that is the very essence of their concern)—they simply find stalls offensive from their own personal perspective and would like to see them abolished. No amount of scientific evidence would ever convince the public that such systems are not cruel. (Stookey, personal communication to Rick Klassen and Laurie Connor, February 16, 1983)

I know this to be true from personal experience. In 2007, I met for eight hours with two top executives from pork-producer Smithfield Farms. I summarized some of my arguments about animals' teloi for them and asked that they consider developing one or more group-housing facilities for sows.

Much to my surprise, I received a call from them six months later, telling me that they would be converting *all* their North American facilities to group housing. Showing great prescience and sensitivity to consumers, they had instituted surveys and focus groups for their customers. I had told them that 75 percent of the public found sow stalls unacceptable. They gently corrected me and told me that it was in fact 78 percent. They are well on their way to achieving their goal. Smithfield has thus taken a major step toward respecting swine telos. Unfortunately, the rest of the companies in the US pork industry are less savvy, and they are still spending a great deal of money to defend gestation crates, a battle they clearly will not win. Well over a hundred retailers and restaurant companies now refuse to buy from producers using sow crates.

Aristotle, like the authors of the Bible, believed in natural kinds, essentially in teloi that are fixed and immutable. Now that we believe in evolution by natural selection, does this cast doubt on the concept of telos? We now believe that life is a process of constant flux and ever-present mutation. Aristotle categorically denied such a possibility, on the grounds that if it were the case, nature would be unknowable. (This of course evidences the influence of Plato on Aristotle; Aristotle's natural kinds are essentially Plato's eternal Forms brought into nature.) For us, a species or a telos is a snapshot of the constant change in nature, somewhat artificially frozen in time. From a metaphysical point of view, that represents a huge difference between a modern perspective on telos and an Aristotelian one. But in terms of knowing an animal's nature, the glacial pace of change makes no difference to our understanding until such time as the ever-occurring, minute changes that kinds of organisms undergo combine to create a phenotypic new kind. Even as animal natures change imperceptibly, we are perfectly capable of understanding the animals' needs and natures flowing from their teloi at any given historical moment.

Does the fact that certain kinds of animals have had their teloi shaped and modified by domestication make any difference to our argument? I do not believe so. Telos is telos regardless of whether it has developed strictly by natural selection or by conscious and deliberate modification by selective breeding. All farm animals and companion animals, what are generally referred to as "domestic animals," have had their teloi shaped by humans in the course of deliberate breeding. Selecting for relative docility among cattle; the willingness to restrict their extent of roaming and geographical range in dogs; tractability for riding and hauling in equids: all represent major steps in domestication. An excellent example of this can be found in looking at donkeys versus zebras. Though the two species are closely related and can interbreed, producing fertile offspring, their teloi are very different. To my knowledge, it is virtually impossible to break and train a zebra to be saddled and ridden. Similarly with dogs and wolves. The two species can interbreed, producing fertile offspring, but the wolf will not restrict its range and cannot be made into a predictable house pet.

In general, it is probably easier to know the full teloi of domestic animals than those of "wild," or undomesticated, animals, in that we have partly consciously shaped the former to suit our needs. And since we live with them, we are likelier to understand them better. I recall a cattle veterinarian once telling me that, as a lifelong cattleman, he could certainly write a book explaining "the mind of the cow" were he not fearful of losing credibility with his peer group, all imbued with the standard scientistic skepticism about the reality and knowability of animal mind. But this is not to say that the teloi of nondomestic animals are in principle unknowable or even conceptually more difficult to know. For scientists like Jane Goodall, who devote many years to the study of the telos and mind of a kind of "wild" animal, in her case the chimpanzee, those teloi become every bit as knowable as those of more familiar animals. In both cases, what is required is careful, sympathetic, and unobtrusive study of the animal over long periods of time. As a matter of fact, just because we live with domestic animals does not mean we understand their natures—witness dog owners who believe that a wagging tail is an indicator of friendliness rather than excitement.

The End of Husbandry

EARLIER IN THIS book, we discussed the "ancient contract" humans histor-ically created with farm animals and the extent to which this assured good husbandry, good care, and good welfare for farm animals insofar as we (as producers) did well if and only if the animals did well. If our analysis of the decline of husbandry in the face of the industrialization of agriculture has been correct, it follows that all farm animals raised under industrial con-ditions, with no attention paid to their needs, natures, or teloi, will suffer from a considerable impairment of welfare. In fact, it is only beef cattle (and perhaps sheep) raised under extensive pastoral conditions (i.e., open range) whose telos is respected in modern agriculture. This is not to say that there is no infringement on animal welfare in extensive cattle production. So-called management procedures, such as castration, branding, and dehorning, all very painful, are accomplished without pain control and thus have major impact on the animals' welfare that should not be underestimated, avoidance of pain of course being part of animal telos, but at least extensive cattle pro-duction does not involve systematic violation of telos.

After World War II, the ancient husbandry contract with animals was broken by humans. Tellingly, at universities, departments of "animal hus-bandry" became departments of "animal sciences," their academic goals newly defined not as care but as "the application of industrial methods to the production of animals so as to increase efficiency and productivity." With "technological sanders," as I call them—hormones, vaccines, antibiotics, air-handling systems, mechanization—we could force square pegs into round holes and place animals into environments where they suffered in ways irrel-evant to productivity. If a nineteenth-century agriculturalist had tried to put

100,000 egg-laying hens in cages in a building, they all would have died of disease in a month; today such systems dominate.

The new approach to animal agriculture was not the result of cruelty, bad character, or even insensitivity. It developed rather out of perfectly decent, prima facie plausible motives that were a product of dramatic and significant historical and social upheavals that occurred after World War II. At that point in time, agricultural scientists and government officials became extremely concerned about supplying the public with cheap and plentiful food for a variety of reasons. In the first place, after the Dust Bowl and the Great Depression, many people in the United States had soured on farming. Second, reasonable predictions of urban and suburban encroachment on agricultural land were being made, with a resultant diminution of land for food production. Third, many farm people had been sent to both foreign and domestic urban centers during the war, which created in them a reluctance to return to rural areas that lacked excitement; recall the post–World War I song "How Ya Gonna Keep 'Em Down on the Farm (After They've Seen Paree)?" Fourth, having experienced the spectre of literal starvation during the Great Depression, the American consumer was, for the first time in history, fearful of an insufficient food supply. Fifth, projection of major population increases further fueled this concern.

When the above considerations of loss of land and diminution of agricultural labor are coupled with the rapid development of a variety of technological modalities relevant to agriculture during and after World War II and with the burgeoning belief in technologically based economics of scale, it was probably inevitable that animal agriculture would become subject to industrialization. This was a major departure from traditional agriculture and a fundamental change in agricultural core values—the industrial values of efficiency and productivity replaced and eclipsed the traditional values of "way of life" and husbandry.

Between World War II and the mid-1970s, agricultural productivity—including animal products—increased dramatically. In the hundred years between 1820 and 1920, agricultural productivity doubled. After that, productivity continued to double in ever-decreasing time periods. The next doubling took thirty years (between 1920 and 1950); the subsequent doubling took fifteen years (1950–1965); and the next one took only ten years (1965–1975). As R. E. Taylor (1984) points out, the most dramatic change took place after World War II, when productivity increased more than fivefold in thirty years. Fewer workers were producing far more food. Just before World War II, 24 percent of the US population was involved in production agriculture; today the figure is well under 2 percent. Whereas in 1940 each farm worker supplied food for eleven persons in the general population, by 1990 each farm worker was supplying eighty persons. At the same time, the proportion of disposable income that people spent on food dropped significantly, from 30 percent in 1950 to 11.8 percent in 1990. Today it is 7 percent.

There is thus no question that industrialized agriculture, including animal agriculture, is responsible for greatly increased productivity. It is equally clear that the husbandry associated with traditional agriculture has changed significantly as a result of industrialization. One of my colleagues, a cow-calf cattle specialist, says that the worst thing that ever happened to his department is betokened by the name change from the Department of Animal Husbandry to the Department of Animal Sciences. No husbandry person would ever dream of feeding cement dust, sheep meal, or poultry waste to cattle, but such "innovations" are entailed by an industrial/efficiency mind-set.

For our purposes, several aspects of technological agriculture must be noted. In the first place, as just mentioned, the number of workers has declined significantly, yet the number of animals produced has increased. This has been possible because of mechanization, technological advancement, and the consequent capability of confining large numbers of animals in highly capitalized facilities. Of necessity, less attention is paid to individual animals. Second, technological innovations have allowed us to alter the environments in which animals are kept. Whereas in traditional agriculture, animals had to be kept in environments for which they had evolved, we can now keep them in environments that are contrary to their natures but congenial to increased productivity. Battery cages, into which egg-laying hens are crowded, unable even to stretch their wings, and gestation crates confining pregnant sows provide examples of this point. The friction thus engendered is controlled by technology. For example, crowding of poultry would once have been impossible because of flock decimation by disease; now antibiotics and vaccines allow producers to avoid this self-destructive consequence.

As we mentioned earlier, only the extensive, pasture–based beef industry has retained the traditional ethic of animal husbandry. Some years ago, over two months, I talked to a half dozen rancher friends of mine. Every single one had experienced trouble with scours, a serious diarrhea condition, in their calves, and every one had spent more on treating the disease than was economically justified by the calves' monetary value. When I asked these men why they were being what an economist would term "economically irrational," they were quite adamant in their response: "It's part of my bargain with the animal; part of caring for them," one of them said. It is, of course, the same ethical outlook that leads ranch wives to sit up all night with sick marginal calves, sometimes for days in a row. If the issues were strictly economic, these people would hardly be valuing their time at fifty cents per hour—including their sleep time.

Consider the story told to me by one of my colleagues in animal sciences at Colorado State University. This man told of his son-in-law, who had grown up on a ranch but could not return to it after college because it could not support him and all his siblings. (Notably, the average net annual income of a Front Range—that is, eastern slope of the Rocky Mountains—rancher in Colorado, Wyoming, or Montana is about $35,000.) He reluctantly took a job managing a barn of feeder pigs, that is, young pigs being fed till they reached market weight, at a factory farm.

One day he reported a disease that had struck his piglets to his boss. "I have bad news and good news," he reported. "The bad news is that the piglets are sick. The good news is that they can be treated economically." "No," said the boss. "We don't treat! We euthanize!" He proceeded to demonstrate by dashing the baby pigs' heads on the side of the concrete pen and then throwing the still-twitching piglets into a garbage heap. The young man could not accept this. He bought the medicine with his own money, clocked in on his day off, and treated the animals. They recovered, and he told the boss. The boss's response was, "You're fired!" The young man pointed out that he had treated them with his own time and money, and was thus not subject to firing. He did, however, receive a reprimand in his file. Six months later he quit and became an electrician. He wrote to his father-in-law: "I know you're disappointed that I left agriculture, Dad. *But this ain't agriculture!*"

From an ethical perspective, animal welfare was a major casualty following from the industrialization of agriculture. The most obvious aspect of harm to animals from the new industrial agriculture is the patent manner in which it fails to accommodate the needs and interests emerging from animal telos. With animals confined for efficiency and away from forage, much research has been directed toward finding cheap sources of nutrition, which in turn has led to feeding such deviant items to animals as poultry and cattle manure, cement dust, newspaper, and, most egregiously, bone meal to herbivores, the latter of which created BSE, or "mad cow disease." Animals are now kept under conditions alien to their natural needs for the sake of productivity.

Whereas husbandry animal agriculture stressed putting square pegs into square holes and round pegs into round holes while producing as little friction as possible, industrialized animal agriculture forces square pegs into round holes by utilizing technological sanders such as antibiotics, hormones, vaccines, extreme genetic selection, air-handling systems, artificial cooling systems, and artificial insemination to force animals into unnatural conditions while they nonetheless remain productive.

Consider, for example, the egg industry, one of the first areas of agriculture to experience industrialization. Traditionally, chickens ran free in barnyards, able to live off the land by foraging and to express their natural behaviors of moving freely, building nests, dust-bathing, escaping from more aggressive animals, defecating away from their nests, and, in general, fulfilling their telos as chickens. Industrialization of the egg industry, on the other hand, meant placing the chickens in small cages, in some systems placing six birds in a tiny wire cage so that one animal has to stand on top of the others and none can perform any of their inherent behaviors; they are unable even to stretch their wings. In the absence of space to establish a dominance hierarchy, or pecking order, they cannibalize each other, so they must be "debeaked," which produces painful neuromas since the beak is replete with nerves. The animal is now an inexpensive cog in a machine, part of a factory, and the cheapest part at that, and thus totally expendable. Some genetic lines of pigs and chickens are so highly selected for egg and meat production that they have less disease resistance. Pigs have become so susceptible to disease that some farmers have installed antibacterial filters to take germs out of the air that enters the building. This is a technological sander taken to the extreme.

The steady-state, enduring balance of humans, animals, and land has been lost. Putting chickens in cages and cages in an environmentally controlled building requires large amounts of capital and energy and technological "fixes," for example, running exhaust fans to prevent lethal build-up of ammonia. The value of each chicken is negligible, so one needs more chickens; chickens are cheap, cages are expensive, so one crowds as many chickens into cages as is physically possible. The vast concentration of chickens requires huge amounts of antibiotics and other drugs to prevent wildfire spread of disease in the overcrowded conditions. Breeding of animals is oriented solely toward productivity, and genetic diversity—a safety net allowing response to unforeseen changes—is lost. Bill Muir, a genetics specialist at Purdue University, found that commercial lines of poultry have lost 90 percent of their genetic diversity compared to noncommercial poultry, which concerns him greatly (Lundeen, 2008). Small poultry producers, unable to afford the capital requirements, are lost; agriculture as a way of life as well as a way of making a living is lost; small farmers, who Jefferson argued are the backbone of society, have been superseded by large corporate aggregates. Giant corporate entities, vertically integrated, are favored. Manure becomes a problem for disposal and a pollutant, instead of fertilizer for pastures. Local wisdom and know-how essential to husbandry is lost; what "intelligence" there is is hard-wired into the production system. Food safety suffers from the proliferation of drugs and chemicals, and widespread use of antimicrobials to control pathogens in effect serves to breed—select for—antibiotic-resistant pathogens as susceptible ones are killed off. Above all, the system is not balanced—constant inputs are needed to keep it running, to manage the wastes it produces, and to create the drugs and chemicals it consumes. And the animals live miserable lives, for productivity has been severed from well-being.

One encounters the same dismal situation for animals in all areas of industrialized animal agriculture. Consider, for example, the dairy industry, once viewed as the paradigm case of bucolic, sustainable animal agriculture, with animals grazing on pasture, giving milk, and fertilizing the soil for continued pasture with their manure. Though the industry wishes consumers to believe that this situation still obtains—the California dairy industry ran advertisements proclaiming that California cheese comes from "happy cows" and showing the cows on pastures—the truth is radically different. The vast majority of California dairy cattle spend their lives on dirt and concrete and in fact never see a blade of pasture grass, let alone consume it. So outrageous was this duplicity that the dairy association was sued for false advertising, and a friend of mine, a dairy practitioner for thirty-five years, was very outspoken against such an "outrageous lie."

In actual fact, the life of dairy cattle is not a pleasant one. In a problem ubiquitous across contemporary agriculture, animals have been single-mindedly bred for productivity—in the case of dairy cattle, for milk production. Today's dairy cow produces over four times more milk than a cow living sixty years ago. In 1957, the average dairy cow produced between 5,000 and 6,000 pounds of milk per lactation, that is, the period of time the cow produced milk before needing to give birth to a new calf. Nearly sixty years later, it is over 22,000 pounds. From 1995 to 2004 alone, milk production per cow increased 16 percent. From 2005 to 2014, average milk production increased from 19,000 pounds to close to 22,500 pounds (United States Department of Agriculture–National Agricultural Statistics Service, 2015). The result is a milkbag on legs, and unstable legs at that. A high percentage of the US dairy herd is chronically lame (some estimates range as high as 30 percent), and these cows also suffer serious reproductive problems. Whereas, in traditional agriculture, a milk cow could remain productive for ten, fifteen, or even twenty years, today's cow lasts no longer than two lactations, a result of metabolic burnout and the quest for ever-increasingly productive animals, hastened in the United States by the use of the hormone bovine somatotropin (BST) to further increase production. Such unnaturally productive animals naturally suffer from mastitis, or teat infection, and the industry's response to mastitis in portions of the United States has created the new welfare problem of docking cow tails without anesthesia in a futile effort to minimize teat contamination by manure. Still practiced, this procedure has been definitively demonstrated not to be relevant to mastitis control or lowering somatic cell count (Stull, Payne, Berry, and Hullinger, 2002). (In my view, the stress and pain of tail amputation, coupled with the concomitant inability to chase away flies, may well dispose to more mastitis.) Calves are removed from mothers shortly after birth, before receiving colostrum, creating significant distress in both mothers and infants. Bull calves may be shipped to slaughter or a feed lot immediately after birth, generating stress and fear.

The intensive swine industry, which through a handful of companies is responsible for 90 percent of the pork produced in the United States, is also responsible for significant suffering that did not affect husbandry-reared swine. We have already examined the most egregious practice in the confinement swine industry and possibly, given the intelligence of pigs, in all of animal agriculture, the housing of pregnant sows in gestation crates or stalls—essentially, very small cages. The *recommended* size for such stalls, in which the sow spends her entire productive life of about four years, with a brief—though no better—exception we will detail shortly, according to the industry is three feet high by two feet wide by seven feet long—this for an animal that may weigh 600 or more pounds. (In reality many stalls are even smaller.) The sow cannot turn around, walk, or even scratch her rump. In the case of a large sow, she cannot even lie flat but must remain lying on her sternum. The exception alluded to is the period of farrowing, or birthing—approximately three weeks—when she is transferred to a "farrowing crate" to give birth and nurse her piglets. The space for her in this cage is not greater, but there is a "creep rail" surrounding her so the piglets can nurse without being crushed by her postural adjustments. We have already examined the extent to which intensive swine production makes a mockery of the needs and interests determined by swine telos.

All other areas of agricultural confinement similarly violate animal telos. Another terrible example is provided by the confinement veal industry. Originally intended to provide a market for unwanted bull calves coming from the dairy industry, the veal industry catered to a market demanding pale, soft, tender "milk-fed" flesh. This was accomplished by keeping the calves in dark, solitary confinement (calves are by nature social), keeping them borderline anemic (iron deficient), and preventing them from exercising (to keep the flesh tender), so that they often literally need to be dragged from pen to truck, with their muscles so underdeveloped that they cannot walk. Significantly, I have never met anyone from a cowboy background who will eat veal so produced. As one cowboy said to me, "If people want to eat veal, we can kill some calves—we don't have to torture them!" Fortunately, public animal-welfare pressure has forced an end to this method of production, with all US veal producers as of 2012 committed to raising calves in group housing.

Telos accommodation is presuppositional to an acceptable system, not a negotiable point. Using the category of telos and its violation to assess animal welfare shows us the stunning number of ways that industrial agriculture hurts the animals being raised under its aegis, sometimes in ways for which we do not have words. Obvious examples of abuse such as keeping social animals isolated, active animals immobilized, and grazing animals away from pasture surely cause significantly negative experiences for the animals even if we do not name them. Separating infants from mothers at one day of age is monstrous—dairy cows subjected to such separation may bawl for weeks. Simply providing feed for animals whose nature is to forage or otherwise work for it is no favor—given a choice, even chickens will work for food. We even control and abort animals' sexual and reproductive lives, by utilizing artificial insemination and sexual neutering.

Equally reprehensible is the way in which we mutilate farm animals for our convenience and to enable them to fit into the highly artificial living conditions. In a 2012 paper written for the World Society for the Protection of Animals with Ian Duncan, I chronicled many of these mutilations, from branding, castration, and dehorning to countless others, normally performed without anesthesia or analgesia. There are *no* drugs federally approved for analgesia in farm animals because of fear of residues in carcasses or animal products. Following are descriptions of those mutilations, by industry.

Several mutilations of animals are commonly practiced in the beef industry. Branding of cattle by the use of a hot iron to create an indelible mark on the skin by infliction of a third-degree burn is extremely painful and works by destroying melanocytes, or pigmentation cells. The purpose is to provide proof of ownership, with each ranch employing a unique, centrally registered mark, and to allow for easy recognition of one's cows under mixed-range conditions, where many different animals with numerous different owners may graze together. In addition, ranchers claim that brands help to prevent rustling (that is, theft of cattle). With periodic change in cattle ownership, an animal may be branded more than once.

Historically, there were few alternatives for permanently identifying cattle, nor were there methods for controlling the pain of the burn. Over the past thirty years, attempts have been made to persuade western ranchers that, in today's world, where industrial agriculture has become increasingly less acceptable to society and a return to husbandry agriculture is sought, they would do well to underscore their commitment to animal welfare by eliminating painful management practices and marketing their beef as the humane meat product. A group of us at CSU developed a method of creating, storing, and comparing digitized images of cows' retinas, which have more data points than human fingerprints. Cattlemen could employ other biometric identifiers or electronic forms of identification, such as microchips. All such methods provide permanent, unalterable forms of identification and have the additional advantage of facilitating trace-back in the event of disease outbreak. Conservative ranchers have resisted moving to alternative methods of identification in spite of the overwhelming evidence that hot-iron branding is extremely painful. If asked to justify the infliction of a third-degree burn morally, cowboys will cite the trade-off involved in living extensively in exchange for a short-term burn pain. However, in addition to the cost to the animals in terms of pain, there is also a monetary cost to the industry due to damage to their hides.

Knife castration of beef cattle is another painful management practice originating in antiquity. Typically, neither anesthesia nor analgesia is utilized to control the attendant pain, which has been well documented. Castration is done to reduce aggressiveness in male animals, thereby minimizing aggressive interactions and danger to humans, as well as to prevent unplanned impregnation of female animals and to improve the perceived quality of the meat. Sometimes castration is accomplished by "banding," that is, placing elastic or rubber bands around the testicles, creating ischemia so that the testicles eventually die and shrivel. As a prolonged insult, banding appears to be more painful than knife castration, although bloodless. Some ways of mitigating knife castration include raising and marketing young bulls, which has been done successfully; the use of local anesthetics and subsequent analgesics to mitigate pain; chemical castration (where injections of toxic chemicals or sclerosing agents destroy spermatogenic capability); and immunological castration, which involves using the immune system to interfere with the spermatogenic cascade. Castration is particularly irrational economically, as the anabolic growth promotion of the testicles is often replaced by hormonal implants (growth-promoting hormones), which do not work as well as endogenous testosterone and which tend to be viewed with suspicion by consumers. Castration and tail amputation, again without anesthesia or analgesia, are also routinely performed on sheep and goats, and there is also ample evidence that these are painful procedures, no matter how they are performed.

Another procedure, dehorning, is utilized to prevent injury by horned cattle to each other and to humans. When done to adult animals by cutting or gouging out the horns, the procedure is extremely painful. When done on young calves, so-called disbudding of the horn buttons can be accomplished less traumatically but still painfully by use of caustic paste, electric irons, or cutting. Of course, a simple alternative to dehorning is to genetically introduce the poll, or hornlessness gene, into one's herd.

All the mutilations discussed above are regularly performed across North America. Although these procedures are well established by tradition, most ranchers will admit that they could be eliminated or replaced without any significant structural effect on their industry. In a real sense, technological innovation is quite capable of rendering these mutilations irrelevant.

Over the past four decades, tail amputation, performed without anesthesia or analgesia, has been increasingly practiced in the dairy industry across the world. Although it has been claimed that docking reduces mastitis because the tail acts as a "brush" to spread manure, this has been refuted by scientific research. Such benefit as it might provide could be accomplished by trimming the tail switch, a painless procedure. Tail docking can cause infection, chronic pain, and immunosuppression. It is therefore good to see that the most recent voluntary Dairy Codes of Practice state that dairy cattle should not be tail docked unless medically necessary. However, since adherence to that code is voluntary, many dairymen still continue the practice of tail docking.

The rise of intensive, industrial, high-technology agriculture has created a demand for many more animal mutilations, making it easier for such agriculture to violate animal nature. Whereas the mutilations recounted above are not essential to raising cattle and other animals under extensive conditions, and could theoretically be eliminated, this is far less the case with mutilations called forth by industrial conditions in the poultry and swine industries.

Consider modern egg production. Cannibalism can lead to high rates of mortality in battery chickens kept in restrictive battery cages, and feather pecking causes injury and loss of thermo-regulatory ability. Ironically, the industry labels cannibalism and feather pecking as "vices," as if chickens are morally blameworthy for engaging in such behavior, whereas in reality inappropriate breeding and intensified production have caused that aberrant behavior. The "solution" to this set of problems is a mutilation known as "debeaking," or "beak trimming," wherein the frontal portion of the upper beak is cut off with a hot blade with no anesthesia or analgesia. Although beak trimming as practiced by the industry does not decrease the incidence of these behavior patterns, it does render the beak significantly less effective in producing injury.

For many years, the egg industry argued that beak trimming was a benign procedure, no more invasive or hurtful than cutting nails in humans. However, it is now clear that this is not the case and that trimming causes behavioral and neurophysiological changes betokening both acute and chronic pain, as mentioned earlier. After the chickens' beaks are trimmed with a hot blade, the damaged nerves in their nerve-filled beaks grow randomly and develop into extensive neuromas, known to be painful in both humans and animals. Furthermore, these neuromas show abnormal discharge and neural-response patterns known to be indicative of acute and chronic pain syndromes in mammals. Behavioral and white-cell responses to beak trimming further evidence this conclusion. There is also evidence that the pain of debeaking may ramify and cause pain during eating, resulting in weight loss.

The meat sector of the poultry industry also engages in mutilations. Male chicks (destined to become broiler breeders) and turkeys of both sexes often have a toe amputated in the hatchery to prevent them from injuring other birds. There is evidence that de-toeing, again performed without any anesthesia or analgesia, causes acute pain.

Another mutilation commonly performed in poultry is called "dubbing," which involves removing the comb on top of a male chicken's head, again, without anesthesia. This is done to prevent later injury to the comb and potential infection. In turkeys, surgical removal of the fleshy protuberance above the beak is known as "desnooding," and is again performed without pain control of any kind.

It is essential to emphasize that none of these poultry mutilations would be necessary if the animals were raised under the sorts of conditions they were evolved to cope with. (For example, under extensive circumstances animals can flee more aggressive conspecifics, and therefore a practice like debeaking is unnecessary.) Humans raised poultry for thousands of years without resorting to the procedures allegedly necessitated by confinement agriculture.

Another area of confined-animal production heavily dependent upon mutilation is the swine industry. Young piglets from one to ten days after birth are subjected to a battery of invasive procedures: vaccination, ear notching for identification (in some cases), teeth clipping, tail docking, and castration if male. As usual, pain control is almost never utilized for these procedures in North America, though parts of Europe are now making it mandatory. Producers often argue that these manipulations are minimally invasive, but common sense tells us otherwise, particularly when all these procedures are performed at once. There is also abundant evidence that these mutilations are acutely painful.

Teeth clipping and tail docking are management procedures carried out to solve problems brought about by a combination of severe confinement and intensive genetic selection for fast growth. Piglets' deciduous teeth, also known as "needle teeth," are clipped to prevent laceration of the sows' udders and abrasion of the faces of other piglets during competition for teats. Tail docking was virtually unknown before the development of intensive production but is now routinely done without pain control to prevent tail biting, a behavior pattern that generally increases once begun and spreads to biting other parts of the body. A victim of tail biting gradually ceases to be reactive to being bitten, in something analogous to learned helplessness. Infection often ensues, and it can become systemic.

Pigs are very highly motivated to root and forage for food. When they are kept in confinement systems with a lack of substrate to forage in, this behavior seems to be redirected toward other pigs' tails. Once the chewing of tails causes a bleeding wound, an attraction to blood causes the behavior to escalate. Under extensive conditions they had the space to get away from one another. It is only in confinement that tail biting became a serious problem. The response of producers has been to amputate the distal half of the tail, a surgical solution to a human-induced problem arising from keeping the animals in a pathogenic environment. Once again, as mentioned earlier, tail biting is referred to as a "vice," as if the pig is bad for doing it. Surgical solutions to human-caused animal problems are not morally acceptable. Humans ought to change the environment to a less pathogenic one, not mutilate the animal. Better husbandry, provision of straw, and alleviating boredom can all reduce tail biting.

In sum, confinement agriculture is the veritable poster boy for diminished animal welfare, not only in terms of consistent and flagrant violation of telos but also in terms of infliction of significant uncontrolled pain on farm animals. In all fairness, some of these painful procedures predate the industrialization of agriculture, mostly castration, branding, and dehorning, but at least they were not in addition to the endless numbers of violation of telos accompanying the development of confinement agriculture. The remaining mutilations were created in support of the failure to fit agricultural systems to animal nature.

Additional problems growing out of the industrialization of animal agriculture were introduced to the public in 2008 by the first intensive study of confinement agriculture ever undertaken. Already highly influential in this area, having received more than eight hundred positive editorials across the United States, is the 2008 report of the Pew Commission on Industrial Farm Animal Production, "Putting Meat on the Table," the first systematic exploration of industrial animal agriculture by an independent commission of experts. The commission's report can be seen as articulating society's nascent concerns about industrialized animal agriculture in a variety of areas.

The Pew Commission on Industrial Farm Animal Production began when the Johns Hopkins School of Public Health, the best-funded school of public health in the United States, garnering fully 25 percent of federal research money in public health, was completing a study of water contamination in the Delaware-Maryland-Virginia area, home to a large segment of the poultry industry. Investigators from Hopkins were disturbed when they found cutting-edge human antibiotics residue there, and they reported back to Robert Lawrence, the director of the Hopkins Center for a Livable Future, a unit concerned with health and sustainability. Lawrence successfully petitioned the $6 billion Pew Charitable Trust to fund a study of industrial animal agriculture and issue a report.

The chairman of the commission was the former governor of Kansas John Carlin, a wise politician who was raised on a dairy farm. The remaining commissioners, including myself, were chosen for their knowledge in areas relevant to commission concerns and were acknowledged experts in their fields. This assured our credibility, which we knew the industry would attack. Other commission members were

- Michael Blackwell, former dean of the College of Veterinary Medicine at the University of Tennessee, Knoxville and highest-ranking veterinarian in the US Public Health Service. Expertise: public health, animal disease.
- Brother David Andrews, former executive director of the Catholic Rural Life Conference. Expertise: rural sociology.
- Fedele Bauccio, founder and CEO of Bon Appétit Management, the first food-service company to address issues of food ethics, which caters 250 million meals a year.
- Tom Dempster, South Dakota state senator.
- Dan Glickman, former US secretary of agriculture.
- Alan Goldberg, founding director of the Center for Alternatives to Animal Testing at Johns Hopkins School of Public Health. Expertise: animal welfare.
- John Hatch, professor emeritus of health behavior and health education at the University of North Carolina School of Public Health. Expertise: public health.

- Dan Jackson, rancher and former president of the Montana Stockgrowers Association.
- Frederick Kirschenmann, distinguished fellow, Leopold Center for Sustainable Agriculture, Iowa State University. Expertise: sustainable agriculture.
- James Merchant, dean of the University of Iowa College of Public Health. Expertise: occupational and environmental health, rural health, public health.
- Marion Nestle, Paulette Goddard Professor of Nutrition at New York University and best-selling author on food issues.
- Bill Niman, founder of Niman Ranch, a company supplied by six hundred family farmers producing humane meat.
- Mary Wilson, leading expert in infectious disease at Harvard Medical School and Harvard School of Public Health.

Tom Hayes, senior vice president at Cargill, participated in all our discussions but resigned before the report was released.

This commission met for over two years across the United States and gave five congressional briefings; we released our final report in May 2008. Our group was funded to hire whatever expert consultants were required, liberally used expert witnesses, and established all our final conclusions by consensus, which was not easy but assured a united front.

In our deliberations, we focused on five interconnected problematic areas associated with CAFOs (concentrated animal feeding operations):

1. Antimicrobial resistance: As early as the mid-1940s, the promotion of antibiotic-resistant pathogens was foreseen as a Darwinian consequence of massive, indiscriminate antibiotic use to promote growth, prevent disease, and compensate for poor husbandry. Antimicrobial resistance today has very likely been augmented by antibiotic use in CAFOs, where an estimated 70 percent of the antibiotics now being produced in the United States are used.

2. Environmental despoliation and farm waste: CAFOs produce huge volumes of animal waste that often exceed the capacity of the land to absorb them, especially in inappropriate areas, such as floodplains. CAFOs also pollute air (as in large dairies) and contribute antibiotics, hormones, pesticides, and heavy metals to

water pollution. They utilize huge volumes of fossil fuel and water. Despite all this, they are not regulated as polluting industries.

3. Rural sociology: CAFOs have replaced the independent, self-sufficient family farmer that Jefferson saw as the backbone of American democracy. In barely forty years, the United States has lost over 90 percent of its small hog farmers to a handful of huge corporate entities. This has increased rural poverty and degradation of small communities. Poor people often bear the brunt of CAFO pollution.

4. Other public health issues: High-confinement operations serve as incubators for pathogens, contribute to the antibiotic resistance that reduces our armamentarium against infectious disease, and adversely affect the physical and mental health of people living near them. Workers in CAFOs suffer more health problems and can spread disease in communities.

5. Animal welfare: CAFOs harm most relevant dimensions of animal welfare, from animals' health to their ability to express their natural behaviors, including basic movements such as standing up and turning around or being with others of their own kind. The vast majority of farm animal diseases are "production diseases," that is, diseases that would not be a major problem if animals were raised extensively.

Obviously, all these categories connect to each other (for example, pollution and health). We found that research on the above issues is largely industry sponsored, which naturally skews toward getting results congenial to industry interests.

One of the most significant results of the commission's work was that it exploded the widespread belief that industrial agriculture is the source of cheap food. While indeed creating animal products that are cheap at the cash register, the claim of cheapness excludes what economists call externalization of costs—passing on hidden costs of production to the public. For example, pollution cleanup is passed on to the public, as are the health costs of those living near pathogenic CAFOs: every man, woman, and child living near the megadairies in California's Central Valley spends $1,500 more annually on health care than if the dairies were not there.

The commission concluded with six basic recommendations, which one lobbyist for the industry in personal communication with me called "a blueprint for the future of agriculture":

1. Phase out and ban the nontherapeutic use of antimicrobials.
2. Improve disease tracking by implementing a national animal-identification system.
3. Improve regulation of CAFO waste.
4. Phase out intensive confinement of farm animals within ten years.
5. Increase competition (reduce monopolies) in livestock production.
6. Establish public funding for research on CAFO issues.

What this Pew report has done is show the public the close connection between all these issues. Now environmentalists must be concerned about confinement of animals, and people concerned about rural life and public health must also see the relevance of animal welfare. As one member of the commission said to me on our last day, "I used to think animal welfare was a fringe issue. Thank you for showing me the centrality of animal welfare to everything else." In the fate of the animals we raise is reflected our own fate—as I tell farmers, "The same forces that put animals in tiny boxes also put you in financial boxes."

What can be done to rectify modern animal agriculture? An extremely unlikely but effective fix would be to phase out violation of animal telos as a presupposition in an agricultural-animal bill of rights. Telos should serve as the basis of animal-welfare science and of creating morally acceptable animal use; we know it is possible to have morally acceptable animal use that works. Not only does attention to telos remediate animal-welfare problems, it also alleviates environmental degradation, disease problems, and many other issues plaguing animal agriculture today.

The critique of industrialized animal agriculture in the Pew report clearly articulates the insight that whole-hog (as it were) adoption of an industrial model for agriculture, and the correlative abandonment of time-tested husbandry concepts, is essentially an experiment that is failing. We know it is failing by virtue of the myriad problems that are following in its wake. Husbandry agriculture was successful and sustainable by virtue of working *with* the land and the animals. It is unwise to tinker with a balanced aquarium. Respect for the nature of farm animals and the land assured that no unnoticed problems would spoil the system. Failure to respect the established system assured the inevitable development of the sorts of problems the Pew commission articulated: environmental problems; loss of sustainability; loss of "wisdom of the soil"; loss of "animal-savvy" workers; loss of small, independent producers beholden to no one and in a position to make their own decisions; loss of rural communities; erosion of both worker and animal health; ground, air, and water pollution; and all the remaining problems I have chronicled.

When considered singly, the above panoply of problems seems insurmountable, but I believe we have found the key to reversing them—restoration of husbandry together with restoring time-venerated respect for animal telos. It helps to realize that none of those problems were epidemic during the era of husbandry-based agriculture and its concomitant reliance on respect for telos, since respect for telos and the husbandry it entailed assured that *animal welfare* was respected and accommodated in how animals were kept. There was no need for technological sanders since how the animals were to be kept was dictated by their psychological and physical natures—round pegs were kept solidly ensconced in round holes. It never even occurred to anyone to force animals into living conditions that were inimical to their well-being. To do so would have been to harm oneself and one's own self-interest.

Respect for telos and its cousin, knowledge of and respect for the land, also assured that environmental constraints were respected. There were, doubtless, people who overgrazed a given pastoral area. But that was likely to happen only once since no one wished to risk compromising their livelihood or that of their children. And without high technology to lure people with extravagant promises of far greater rewards and productivity, no one attempted to work against nature or to overstep environmental constraints.

Respect for telos also taught stockmen what kinds of conditions were likely to keep animals healthy. In 2000, I was privileged to serve as a member of a World Health Organization committee on prudent use of antibiotics in animal agriculture. While there, I happened to meet a former Swedish secretary of agriculture, and we shared a fascinating conversation. He told me that many years earlier, the Swedish public had eliminated, by referendum, the growth-promotion use of antibiotics, a practice that is widely believed to promote antibiotic resistance in bacteria as the drug promotes the killing of susceptible bacteria and the resulting niche is colonized by bacteria that harbor resistance to commonly used antibiotics; this resistance represents a huge public health issue in the United States. The minister told of the panic that the cessation of antibiotic use in livestock occasioned in producers, who frantically sought legal substitutes for the drugs. At a producer meeting, one old man reminisced about working with his grandfather in a cattle operation. He recalled that his grandfather used to fastidiously clean his livestock-storage facilities between shipments of animals. Desperate producers seized on this idea in a "This is so crazy it might work" mind-set. Sure enough, it did work. And contrary to dire predictions from producers about the cost of meat being dramatically driven up, with scrupulous cleaning the price of meat went precipitously *down,* for producers no longer needed to pay for antibiotics to compensate for dirty, pathogenic conditions.

Finally, under an agriculture based in respect for telos, the only way an operation could get bigger was to acquire more land. One simply could not squeeze more animals onto a chunk of grazing land—the number of animals an operation could accommodate was fixed by the land's grazing capacity and the amount the animals consumed. There was therefore no way that larger producers could force out smaller producers by forcing more animals onto a pasture, so efficiency could not be increased at the expense of animal telos. If you had more land than your neighbor, you could produce more animals. But that did not make you more efficient.

To put the same point another way: the industrialization of agriculture essentially amounted to the replacement of labor with capital. Without massive amounts of capital, the giant meat-producing factories of today, which outcompete smaller, more labor-intensive operations, could not exist. From historical data derived from the intensification of agriculture, it does not appear that industrialization and mechanization of animal agriculture made good husbandry more easily achievable. Instead, the value of good husbandry was replaced by the rearing of very large numbers of animals, with each individual animal worth less than it would have been under husbandry but with the numbers greatly scaled up. Hence, for example, milk cows were significantly more valuable when they could continue to be milked for fifteen or more years (what one dairy producer opposed to modern industrial dairies calls "marathon cows," as opposed to today's "sprinter cows," which last less than three lactations). But today's industrial producer, while getting significantly less value from each cow, may raise ten or twenty thousand cows in one operation, making his money not as a reward for good husbandry, but by having sufficient capital to raise large numbers of animals, however many "inputs" that requires, such as cheap energy to run machinery.

About a dozen years ago, this point was brought home to me in an unforgettable manner. I received a phone call from a colleague in the CSU Department of Animal Sciences, John Edwards. He told me that he had a very interesting Navajo graduate student whom I needed to meet. She had an undergraduate degree in public health from an Ivy League university and was now getting a PhD from CSU in animal sciences. As we sat together in the coffee room, I queried her about her future plans. I bubbled with enthusiasm. "Wow! A female Indian student with your credentials! You should be thinking about Harvard Medical School! The world is your oyster! What are you going to do after you finish your PhD here?"

"I'm going back to the reservation to help my grandmother raise her sheep," she replied.

"Why," I sputtered, "why waste all that education and not take advantage of your unique position?"

I will never forget her reply. "At CSU, and in the Ivy League, I was taught a physiological explanation for the occasional phenomenon of one of my grandmother's ewes having twin lambs. From my grandmother, I learned a very different explanation. She taught me that it was a reward for good husbandry. I prefer the latter explanation."

"I prefer the latter explanation"! What a perfect, beautiful articulation of the point we have been making. She realized that her grandmother's explanation essentially made for a better world, one she preferred to live in! It is a better world in which husbandry and sustainability reign over productivity and profit at all costs. The animals live decent, natural lives. Animal welfare is integral to the system, not an afterthought to be politically imposed. In remaking such a world, the ancient contract with animals is restored. Those who work with animals possess knowledge *and* wisdom. They work outdoors, in nonpolluted conditions, in ways that do not erode their health—or their spirit. The preservation of environmental health is a presupposition of agriculture, not a struggle.

This in turn leads me to a very radical conclusion. Rather than trying to legislate decent amounts of space for farm animals, regulate chemical runoff and pollution, preserve land and water in a sustainable way through political battles, fight for worker health, and struggle to keep small agriculture alive, why not create one overarching, fundamental principle that would fix most of what we are trying to change? The principle is very simple: all agricultural systems designed to produce animals must respect, not work against respect for, animal telos. The template for this already exists as husbandry agriculture has worked well for over ten thousand years. Mechanized, intensive, industrial agriculture is barely a hundred years old and already threatens animal welfare, human and animal health, rural communities, and small, independent food producers.

Political realities and the vast amounts of money generated by corporate industrial agriculture render this solution little more than a pipe dream. Even to make such a suggestion elicits shrill cries of "Luddite!" "Anarchist!" "Hater of progress!" and even "You don't care about the poor!" for a return to an agriculture based in husbandry and respect for animal telos would certainly reduce available food quantities and raise prices. By the same token, we must consider that industrial agriculture is very likely living on borrowed time, not being inherently sustainable. If, for example, we lose cheap energy, the whole system collapses. At some point, the catastrophic potential inherent in what we take for granted could well be realized, resulting in what could be an insoluble crisis for contemporary civilization. Let the readers judge for themselves, based upon the arguments presented here. I am arguing, quite simply, for a return to principles of husbandry that have solidly proven themselves throughout the unfolding of human civilization.

The overwhelming implausibility of a return to husbandry agriculture renders such a solution inherently unrealistic. Thus one must return to ways of fixing, rather than replacing, the current system. But at the very least, we should be mindful of the need to preserve telos and husbandry as regulatory ideals for fixing a system that is broken on many fronts.

Animal Research and Telos

IT IS CLEAR from our discussion that the loss of respect for telos in animal agriculture was largely motivated by greed—the desire to make more money from the ability to increase productivity of animal products at the expense of animal nature. This should surprise no one, as greed is one of the major driving forces motivating human beings. In particular, this is true regarding short-term windfall profits; we have already seen that one of the consequences of the industrialization of animal agriculture is the loss of sustainability, that is, long-term productivity with available resources. As a general rule, we can project that animal well-being will suffer if it stands in opposition to profitability in agriculture. This is likely to be true in agriculture as well as any other pursuit that involves extracting profitability from animals, be it fishing or the production of pharmaceuticals. It is strange, then, to consider the uses of animals in science, animal research, and testing, where catering to greed is likely to be counterproductive to good science, since good science is likely to require extreme respect for animal telos.

In most real-life situations, ethical imperatives compete with considerations of self-interest, usually to the detriment of ethics. For example, where ethical edicts dictate that we not lie, self-interest pushes in the other direction. The cheating husband will lie through his teeth to escape detection by his wife, even while knowing that cheating and lying are wrong morally.

In the case of the treatment of animals utilized in research, a clear exception to this rule is manifest. The way researchers house and husband laboratory animals is utterly removed from the apparent self-interested goal: the desire to achieve realistic and accurate data. For example, when we fail to control the pain or distress or stress that arises in research, by virtue of how we manage or care for our laboratory animals, we are violating multiple moral requirements to minimize negative states in these animals, yet we are also violating multiple research imperatives requiring the obtaining of accurate physiological and metabolic data.

Scientists who would never dream of dragging a relatively complex piece of equipment such as a microtome or an MRI to work behind their pickup truck on a rough road do far worse to the needs dictated by the natures of laboratory animals. I experienced a remarkable and unforgettable example of this phenomenon in the early 1980s when I was invited to deliver the banquet speech to an international meeting of the Shock Society—MD, PhD, and DVM researchers who study circulatory shock, that is, circulatory collapse, which is the proximal cause of death in all species possessing a circulatory system.

The meeting was held at the Jackson Lake Lodge, a beautiful venue near Jackson, Wyoming, in the Teton Mountains. Like all federal resorts, the lodge was managed by college students and retirees. As soon as I arrived, I was approached by a small group of young researchers, who asked me to try to exert some influence on the leadership of the society, who promulgated an ironclad policy that any papers published in the society's journal could not report on any research where anesthesia was employed. This was their policy, despite the fact that shock research papers often reported on traumatic shock, or circulatory collapse occasioned by trauma. Traumatic shock was created in many cases by insertion of an animal into a Noble-Collip drum, a device that rotated at sixty cycles per minute, creating random trauma by bouncing the animal off random protuberances. The use of analgesia or anesthesia was forbidden, using the excuse that the animals were modeling humans who, for example, were traumatized in automobile accidents and were thus not anesthetized. In fact, a film of such a procedure using dogs was shown at the conference, with the projector being operated by a college student. At the close of the session, the student approached the president of the society and remarked, "We can't wait for you sons of bitches to be out of here," greatly upsetting members of the society.

That evening, I gave my talk. I began by sarcastically "applauding" the scientific spirit pervading shock research, by virtue of not wanting to introduce variables such as anesthesia that create disanalogies to real-life occurrences of shock. I continued by pointing out that the same scientific spirit doubtless led them to control for known variables that deform the shock response. Just that very year, I pointed out, D. Gärtner, in Hanover, Germany, had shown that one could take two groups of standard laboratory rats, keep them under identical conditions for six months, and then one day come in and move one cage three feet. By the time a hundred seconds had gone by, numerous plasma variables would indicate that the rats in the cage that had been moved were exhibiting a microcirculatory shock profile that would persist for forty-five minutes. "Surely," I continued, "you people control for these sorts of variables?" And other variables, I said, that are known to impact on animal physiology, such as the personality of the caretaker? All present indicated that they had never heard of these results. "If that is the case, *how dare you* withhold anesthesia as you do, the provision of which is surely a moral presupposition of the kind of work you do?" Furthermore, I argued, my voice rising, "I will bet you the $200 currently in my wallet against a Coke that we can create a function mapping the physiological responses of an anesthetized animal onto an unanesthetized one!"

Never before nor since have I seen an audience that silent. The embarrassment finally ended when the president of the society, also the journal editor, cleared his throat and said, "I'm glad you brought that issue up. I was going to announce at the end of your talk that our journal will no longer accept studies done on *unanesthetized* animals. This is a radical change in policy." The audience spontaneously began to applaud, particularly the young people.

This is a paradigmatic and dramatic example of both ignorance of, and lack of attention to, the needs and interests of laboratory animals. As I have argued in great detail, and as I recall the British journal *Nature* remarked in the 1960s, it was once believed virtually universally that animal use in science is not a moral issue, it is a scientific necessity. Under the influence of the powerful ideology that pervaded science after Newton, science was seen as "value-free in general, and ethics-free in particular." As we discussed, I have termed this the scientific ideology, or the common sense of science.

Despite the fact that the United States has produced some of the most sophisticated science in human history, as well as truly dramatic technology attendant on that science, the American public is far from unequivocally supportive of science. There are many reasons for that lack of support. Perhaps the most patent reason is the appalling scientific illiteracy and rampant anti-intellectualism that is pervasive throughout American history.

We cannot underestimate the degree of scientific illiteracy in the US public and elsewhere. First described with regard to intellectuals in universities by C. P. Snow in the 1950s as "the two cultures in conflict," that is, science and everything else, there is little reason to believe things have improved. As Keith Black wrote in the *Cedars-Sinai Neurosciences Report* (2004),

> America's best and brightest used to go into science and medicine, but no longer. The United States consistently ranks in low comparison to other developed countries on assessments of scientific literacy. One half of the American public does not know the earth goes around the sun once a year, and believes that the earliest humans lived at the same time as the dinosaurs. . . . A 1996 National Assessment of Educational Progress survey found that 43 percent of high school seniors did not meet the basic standard for scientific knowledge.

Jon Miller of Northwestern University, who studies scientific literacy in the United States, affirms that only 20 to 25 percent of Americans are "scientifically savvy and alert. . . . [The rest] don't have a clue." According to Miller, US adults do not know what molecules are, fewer than a third know that DNA is the key to heredity, and only 10 percent know what radiation is; 16 percent of high school science teachers are Creationists; and two-thirds of the US public wants creation taught along with evolution, according to a 2004 CBS News poll (Dean, 2005).

This should not surprise us, given Richard Hofstadter's Pulitzer Prize–winning 1963 book, *Anti-intellectualism in American Life,* pointing out the deep current of anti-intellectualism in American history going back to the founding of this country. And not only is the United States anti-intellectual, it is openly hostile to science. As the economist Jeffrey Sachs (2008) was quoted as saying in *Business World,* "By anti-intellectualism I mean an aggressively anti-scientific perspective, backed by disdain for those who adhere to science and evidence." Consider also that the use of stem cells and biotechnology have been widely rejected for bad reasons.

Other factors both follow from and augment antiscience feeling. These include the unfortunate resurgence of "magic thinking," reflected in the reappearance of Creationism hostile to evolution; in the billions of dollars spent on evidentially baseless "alternative medicine"; and in the fact that cryptozoology books sell more than all bioscience books combined. The never-ending appeal of the Frankenstein myth as a metaphor for scientific progress in turn ramifies in public skepticism regarding scientific advances.

As mentioned earlier, my work on augmenting the moral status of animals in society began during the 1970s with my realization that animal ethics needed to be based in judo, or recollection, not sumo, as well as with my practical efforts to bring moral concern for animals into veterinary education. Though I was dimly aware of the animal issues raised by confinement agriculture, my first effort at effecting change in animal use was focused upon animal research. This was the case for a number of reasons. First of all, as an intellectual housed in an academic environment, with appointments not only in philosophy but in animal sciences and biomedical sciences, and as someone who was under a National Science Foundation grant to teach philosophical and moral issues in biology, I instinctively believed that scientists would be far more prone to be receptive to ethical arguments than farmers, whose goal was profit. This was before I had begun to understand the pernicious power of scientific ideology thoroughly to block scientists from thinking about ethics in science. I was shocked to learn that scientists were utterly blind to the myriad ethical issues that arose from invasive use of animals in science.

In 1976 this was called to my attention by my co-teacher in veterinary ethics, Harry Gorman, and by the newly arrived director of laboratory animals at Colorado State University, David Neil. Between them, Gorman and Neil enjoyed more than fifty years of experience with research animals, in the United States, Canada, Great Britain, and the military, and both were firmly convinced that research animals were not getting even a semblance of a fair shake. When Gorman arrived at CSU after a distinguished career in the military as an experimental surgeon who in fact invented the artificial hip joint, he was shocked to discover when setting up his orthopedic research lab that the CSU veterinary pharmacy carried no narcotics for analgesia. "What do you need that for?" he was asked. He replied that his research was painful and consequently he needed pain control for the animals. "Oh, just give them an aspirin," he was derisively told.

After extensive discussion of moral obligation to animals used in research, we resolved to press for legislation providing some measure of protection for research animals, for, shockingly, none existed. In the 1970s, there was absolutely no legal protection for animals used in teaching, and research laboratories after World War II were even allowed to obtain animals from pounds to use for experimentation. In general, the biomedical research community had successfully countered any legislative intrusion into the research process from World War II until the 1960s, cannily portraying animal research as a scientific necessity, not an ethical issue, and portraying those who raised moral questions about animal research as misanthropes unconcerned about human health: "animal lovers and people haters."

In the mid-1960s, however, two events took place that made it politically necessary for Congress to address animal research, at least on a superficial level. Here are the incidents, as described in official United States Department of Agriculture (USDA) history:

> In July 1965, a Dalmatian named Pepper disappeared from her backyard and was later spotted by a family member in a photograph of dogs and goats being unloaded from a Pennsylvania animal dealer's truck. The family discovered that Pepper had been sold to a dog dealer in New York State. When the family confronted the dealer, they were refused entry onto the property. . . . Events led to a telephone call to Congressman Resnick's office in the District where the dog dealer was located. However, even Mr. Resnick's intercession failed. Angered by the dealer's refusal to admit the family, Congressman Resnick decided to introduce a bill to prevent such wrongs. Pressure from the Pennsylvania State Police led to an admission that Pepper had actually been sold to a hospital in New York City. In the end, Pepper had been used in an experiment and was euthanized. Pepper's disappearance, however, had galvanized several members of the House and Senate to introduce legislation to prevent future incidents.
>
> Congressman Resnick's bill was introduced in the House and required that dog and cat dealers and the laboratories that purchased them be licensed and inspected by the U.S. Department of Agriculture, and required to adhere to humane standards established by the Secretary of Agriculture. Similar legislation was introduced in the Senate and co-sponsored by Senator Warren Magnuson and Senator Joseph Clark.
>
> According to Magnuson:

The Committee on Commerce opens the first of two days of hearings this morning on a question which is of very great concern to millions of Americans: The protection of the pet owner from loss of the pet through theft and the assurance that animals in the hands of dealers will be humanely treated.

I would like to emphasize that the issue before us today is not the merits or demerits of animal research. We are interested in curbing petnapping, catnapping, dognapping, and protecting animals destined for research laboratories, while they are in commerce. We are not considering curbing medical research.

I have always considered myself a friend of the medical researcher. Yet, we do not think we can allow the needs of research, great as they may be, to promote either the theft of a child's pet or the growth of unscrupulous animal dealers.

From their introduction, both bills faced opposition. However another event was about to occur that would make it harder for the legislation to fail. While hearings on the House bill were being held by Congressman W. R. Poage, Chairman, House Agriculture Committee, an article appeared in *Life* magazine with photographs taken by Stan Wayman during a raid by the Maryland State Police, documenting the abuse of dogs in a dealer's facility. The resulting public outcry led opponents of the legislation to modify their stand and to attempt to seek exemptions for research facilities rather than to push for complete defeat of the legislation. Although both the House and Senate bills were initially weakened by exemption for laboratories, Senator Mike Monroney prepared an amendment that restored coverage of laboratory animals. Despite attempts made to defeat this amendment, newspapers throughout the country offered editorial support for the Monroney amendment. In the end, the Senate Commerce Committee bill was passed by the Senate and sent to President Lyndon Johnson who signed the bill into law on August 24, 1966. The bill became Public Law 89–544. (USDA–Animal and Plant Health Inspection Service, "Legislative History of the Animal Welfare Act," www.aphis.usda .gov/ac/awahistory.html [no longer available])

It is absolutely essential to note that we are not even close to dealing with rational animal ethics per se in this legislation. The unabashed reasons for these laws are protection of human sensibilities—that is, concern that people's beloved possessions, their pets, not be dognapped or catnapped and end up in experiments—and calming public hysteria. Furthermore, Wayman's photographs struck at the heart of American's love for dogs. In particular, his stark nighttime photo of an emaciated English pointer, little more than a bag of bones, held by a dealer, was bound to galvanize a major emotional response.

When one looks at the Laboratory Animal Welfare Act from the point of view of rational ethical content, one is appalled. As I tell my students, if I were to assign writing a law to a freshman class in animal ethics and receive this 1966 document, I would unhesitatingly fail the students who wrote it. First of all, the act defines "animals" in research as "live and dead dogs, cats, monkeys (nonhuman primate animals), guinea pigs, hamsters, and rabbits." Specifically excluded under the regulations are rats, mice, birds, farm animals, and horses used for food and fiber research. Given that rats and mice were estimated to comprise over 90 percent of the animals used in research, this was hardly a comprehensive research-animal welfare act.

In addition, the regulations state that "animal" shall mean, in addition to the animals listed above, any "other such warm-blooded animal as the Secretary [of Agriculture] determines is being used, or is intended for use, for research, testing, experimentation . . . " The absurdity is blatant. The law authorizes the secretary to determine (that is, find out) which animals are used for research and cover them, yet also to decide, as in the regulations, not to cover certain animals that are in fact so used.

As it turned out, not surprisingly, the animals covered were ones that aesthetically appeal to members of the public. As one USDA inspector said to me in the 1970s, he could bring charges against a researcher or dealer who "abuses" a dead dog yet is powerless against a scientist who is biting the heads off mice and spitting them into garbage cans.

In tandem with this most ethically unsound definition of "animals" came a very restricted notion of the scope of the act:

> The Secretary [of Agriculture] shall establish and promulgate standards to govern the humane handling, care, treatment and transportation of animals by dealers and research facilities. Such standards shall include minimum requirements with respect to the

housing, feeding, watering, sanitation, ventilation, shelter from extremes of weather and temperature, separation by species, and adequate veterinary care. The foregoing shall not be construed as authorizing the Secretary to prescribe standards for the handling, care, or treatment of animals during actual research or experimentation by [a] research facility as determined by such research facility.

In other words, the act was intended to assure research-animal welfare without prescribing standards for "handling, care, or treatment of animals during actual research or experimentation." This is relevantly analogous to a sex manual that covers cohabitation and foreplay but disavows concern with anything having to do with sexual intercourse. In 1970, the act was amended (and renamed the Animal Welfare Act, P.L. 91-579) to include assurance of proper use of anesthesia, analgesics, and tranquilization by the research facility during experiments. However, the absurdity therein was that the regulatory requirement could be met by the research facility's affirming in its annual report that it saw no need for anesthesia, analgesics, or tranquilization, despite its in fact performing painful research.

Our legislation-writing group saw its task as creating moral checks on animal use in research, the most important of which was the legally mandated control of pain, if animals were to be used in ways causing pain. Further, the approach to legislation seemed to follow logically and in accord with common sense: if animals were not getting the best treatment possible consistent with their use in research, indeed if poor treatment was sometimes even compromising research by introducing uncontrolled stress and pain variables, we had a societal opening to pass between the horns of (1) those in the research community who insisted on no constraints on the use of animals in research and (2) those animal advocates who would forbid research altogether. Society as a whole, we surmised, was in the middle on those matters.

From 1976 to 1980 we drafted this model legislation until finally we came before the Colorado legislature, where we were quickly demolished: our proposed legislation never made it out of the agriculture committee. In retrospect, we realized our naiveté when we were approached by US Representative Pat Schroeder from Colorado, who pointed out that such legislation could not possibly work in only one state, but needed to be federal, and offered to carry it up for us in Congress.

We learned much in the ensuing five years until our bill passed. In the first place, and contrary to our expectations, we learned that the research community absolutely and completely opposed any legislative assurance of proper animal treatment. There had already been a long tradition in the medical research community of seeing anyone raising questions about the ethics of animal use as "anti-science, anti-human, anti-progress, anti-vivisectionist." Thus, for outlining the legislation in a book, I was called an apologist for the Nazis and the lab-trashers by a reviewer in the *New England Journal of Medicine.* It was as if, to the scientific community, we were hell-bent on stopping medical progress. It was only gradually that I learned that scientific thought was guided by the powerful and immovable ideology declaring science to be "value-free" in general and "ethics-free" in particular, and that this ideology required agnosticism about animal thought, feeling, consciousness, and pain as empirically unknowable. Ironically, opponents of animal research attacked us equally vigorously, in one paradigm case calling me a sell-out for "accepting the reality of science." The latter claim stems from a radical viewpoint according to which animal abuse cannot be remedied incrementally, but requires revolutionary change. Famed activist Henry Spira, though an abolitionist vis-à-vis animal use, nonetheless pointed out that all social revolution in the history of the United States has been incremental. We were quite secure in our belief that we had seized ground that society in general would support, namely, assuring more humane treatment for laboratory animals. Nonetheless, we were told explicitly that we had an uphill battle, if only because the medical community had such a powerful lobby opposed to us. We needed to justify, persuasively and painstakingly, every provision we proposed to make mandatory.

At the root of our legislation was control of pain and distress, the latter encompassing such negative emotional states as fear, loneliness, and boredom. We also mandated acquiescence to the concept that if a procedure would hurt humans, it could be presumed to hurt animals. The research community outrageously claimed to control pain already. We proved this false by doing a literature search on animal analgesia in 1982, when I went before Congress. The result? There were only two papers in scientific journals, one of which said that "there ought to be papers." Such evidence of the neglect of pain control could not be ignored. (That the law has been effective in increasing awareness of the need for analgesia is attested to by the fact that, when I redid the same search about three decades later, I found almost 13,000 published papers.)

Second, in addition to mandating control of pain and distress, the legislation we proposed was intended to break the hold of agnosticism about ethics and mental states among scientists. We did this by requiring institutional animal care and use committees, now known by the familiar acronym IACUCs, which would include both scientists and nonscientists, to review prospectively all protocols and discuss them in terms of proper numbers of animals (allowing neither too many nor too few), pain and suffering, experimental design, species, and so on. Such mandated discussion, we felt, would help the ideology crumble—and it did. Committees also reviewed all teaching and inspected facilities and protocols, that is, the proposed plans for research and teaching. I have been on CSU's IACUC with other senior faculty since 1980.

Third, we proposed in our bill that *all* laboratory animals (including rats and mice, historically excluded from the Animal Welfare Act and still excluded thanks to efforts of the research community) be housed and kept in ways that meet their biological and psychological needs and natures. Unfortunately, Congress was unwilling to grant this, instead mandating only exercise for dogs and the provision of environments for nonhuman primates that "enhances their psychological well-being."

Other provisions included the following: No paralytic drugs were to be used without anesthetic. Multiple surgeries were prohibited unless justified to test a single hypothesis. An animal-welfare information service was to be established at the National Agricultural Library. Research facilities were required to institute and oversee training for researchers and staff on humane practices in experimentation.

In addition, the USDA (which enforced the Animal Welfare Act and these amendments) was to share efforts with the National Institutes of Health (NIH), which, though historically, beginning in the 1960s, having good guidelines for proper care and use of laboratory animals, failed to enforce them. The NIH guidelines were in fact turned into law at the same time as our amendments, and both went into effect in 1987.

Virtually all vertebrate animals used in research were covered by one or another of these new laws, though in a very reactionary move, when the USDA was later planning to include rats and mice under the Animal Welfare Act, the National Association for Biomedical Research, the chief biomedical research lobby group, convinced Senator Jesse Helms to sponsor legislation in 2002 declaring rats and mice not to be animals for the purposes of the act. Gratifyingly, by then such a move was not very popular with the scientific community, as many scientists felt it made them look ridiculous in the eyes of the public. Nevertheless, it prevailed.

The laws, and particularly the regulations interpreting them, established by the USDA's Animal and Plant Health Inspection Service and to a lesser degree by the NIH, are far more complicated than this thumbnail sketch might suggest. For example, there are detailed rules concerning surgery, veterinary care, psychological well-being, and so forth. But, conceptually at least, we now know enough to understand why these laws are indeed revolutionary breakthroughs in animal ethics.

In the first place, some of the above-mentioned absurdities manifest in the 1966 laws have been largely corrected. Although the 1966 claim disavowing any legal control over the actual conduct of research still remains, the procedures mandated clearly belie that claim. Similarly, although the Animal Welfare Act amendment and the Helms law deny legal protection to rats and mice, the NIH law covering all federally funded institutions does cover them in most settings, though there are still some exempted contexts. Farm animals used in biomedicine are clearly covered, and many IACUCs demand biomedical-level control of pain even in agricultural research. Many committees have also applied pain-control rules to invertebrates like the cephalopods, for example, the octopus, where there is excellent scientific reason to believe pain and even distress are present.

Additionally, the laws have significantly eroded the ideology that creates a radical cleavage between ethics and science. Protocol review is inherently replete with substantial ethical discussion, which inevitably has become more and more sophisticated with time. When my own institutional committee began in 1980 (voluntarily, to show Congress such a system could work), we might cover twenty protocols in a ninety-minute meeting, including time to eat lunch and schmooze. Now, judging the same number of protocols consumes three to four hours, and a single controversial protocol can cover an entire meeting. Moreover, scientists on committees understand that the current system is their last chance at self-regulation (as opposed to external regulation) and that the loss of federal funds for their whole institution can be a penalty for not obeying the law. The result is more and more scientists taking animal care and use issues very seriously, and growing committee hostility to that handful of researchers who try to get around the system. Colleagues at the NIH even told me that within five years after the law went into effect in 1987, some committees were discussing cost/benefit issues in terms of animal suffering even though the law does not mandate such discussions. Though most protocols are not rejected, many are modified to the benefit of animals. The biggest problem remaining with the laws is that primacy is still given to the science being done, not to animal welfare. We shall discuss movement in the direction of rectifying this imbalance shortly.

It is also obvious that, from their inception, the laws have eroded the ideological denial of animal mentation, particularly pain. Given that knowledge of and concern for animal pain was almost nonexistent when the laws passed, the USDA wisely concentrated on enforcing control of pain. It was only in the mid-1990s, once pain control was solidly established, the vast majority of young scientists and graduate students had begun to acknowledge pain in animals as axiomatic, and the literature on animal pain had become vast (it is still increasing geometrically), that the USDA mentioned—as a word to the wise—that it would start monitoring "distress," even though, as was the case initially regarding pain, people were still stumbling in the dark with it. Even today, there is little focus by the USDA on distress—shades of scientific ideology.

In drafting this legislation, our group was adamant that the role of the law should be analogous to what Wittgenstein said of his philosophy—it should be an educational ladder that allows or rather compels scientists to transcend their previous agnostic position regarding the ethics of research and the pain and distress of animals and to negotiate routinely in what was historically terra incognita. Given that the law has been in effect for less than thirty years, our goal seems well on the way to being achieved. Ultimately, we hoped to produce a generation of scientists to whom what are now legal stipulations would be second nature and who would have reappropriated common sense and common decency.

Some years ago, at a convention of the American College of Laboratory Medicine where I gave the keynote address, I debated a famous scientist who argued that these laws did not work. His evidence? Some of his own protocols had been turned down by his institutional committee as involving too much pain, so he was not able to proceed with some of his research. He seemed adamant on this point. Seven years later, I encountered the laboratory-animal veterinarian charged with ensuring compliance with the law at his facility, who informed me that the researcher in question now saw the law as essential to scientific activity and as reflecting legitimate social concern that needs to be respected. He is correct. The social furor of distrust that reflected public distrust of scientists' treatment of animals in the 1970s and early 1980s has diminished. There is no question that the revelations of People for the Ethical Treatment of Animals (PETA) and other activists about researcher Edward Taub's mistreatment of baboons (1981) that led to his indictment for cruelty, the atrocities at the University of Pennsylvania head-injury laboratory (1984), and the misconduct revealed at the research portion of the City of Hope (1985) all fueled the passage of our laws. The groundbreaking and unforgettable film *Unnecessary Fuss* (1984), produced by PETA from footage taken by the researchers at the University of Pennsylvania themselves, is considered a major impetus for passage of our law.

Indeed, US Representative Henry Waxman, from California, told me when I testified before his committee in 1982 that even though our bill would be defeated that year in deference to the medical research lobby, it would most certainly pass within two or three years, because rarely did Congress experience such univocal public concern about an issue. And, correlatively, it is clear that the passage of the law blunted constant media coverage of research atrocities and animal misuse, though occasional flares still emerge, particularly in the area of primate use.

Many animal-welfare advocates press for alternatives to the use of animals in research; however, the law requires only that researchers search for alternatives. To use the famous language of William Russell and Rex Burch regarding alternatives to animal use, we may recognize three alternative approaches: replacement of animals by nonanimals, reduction in numbers of animals used, and refinement of animal use. In the short run, the laws have most affected *reduction*, by focusing researchers' attention on previously neglected statistical precision. To be fair, though, committees sometimes see the animal sample as statistically inadequate and need to demand more animals since using too few would invalidate the results and waste those animals being used. One member of our IACUC at CSU articulated the ironic "law of animal numbers," wherein our committee often mandated reducing the numbers of animals requested by researchers when their protocols involved animals that were cheap to purchase and care for, such as rats and mice. Conversely, the committee sometimes had to require increasing, for statistical reliability, the numbers of animals beyond those requested when they were costly to buy and care for—such as horses and cows—even though the research was statistically similar.

Even more than reduction, perhaps, the laws have focused on *refinement* of procedures, notably by demanding timely euthanasia of potentially suffering experimental animals, that is, early "end points," as well as precise end points decided in advance. Unfortunately, end points in infectious-disease research are still very late. Similarly, whereas in the 1970s I saw tumors grown in animals larger than the animals themselves, today tumor size is strictly limited and small.

Animal activists, however, favor *replacement* of animals as the most desirable alternative. Unfortunately, replacement is difficult, requiring both significant money for research and equally significant amounts for validation. Science tends to be conservative, and it demands what it considers full proof of the viability of an alternative before replacing animals as the historical "gold standard," whether they are or not. That is not to say the laws have not encouraged replacement at all. Particularly in teaching, invasive animal use, especially use involving suffering, is much diminished. Whereas once we did multiple-survival surgery and poisoning of animals, even terminal surgeries for teaching are declining, and committees are increasingly asking, "Why can't you film it instead?" A famous example is hemorrhagic shock labs, where medical and veterinary students were forced to bleed animals out and watch the stages to death. Today such labs, once ubiquitous, have been replaced by films or computer simulations in most medical and veterinary schools.

In short, the laws have provided an ongoing mechanism for the scientific community to reflect both on what it does and what society thinks about it in ethical terms. Having said all this, it is necessary to sound a cautionary note and address the fact that the laws are still far from perfect. At best, they represent only the first real steps of ethical evolution. And, although those of us who drafted the laws were above all committed to "enforced self-regulation," as one Australian sociologist put it, inevitably that simple prescription has unfortunately become bloated by relentless bureaucratic proliferation, at both the federal level and the research institution level. When CSU adopted the procedures mandated by our proposed legislation, the system was extremely lean, with one administrative assistant managing the entire program. Today, the regulatory compliance office has grown to more than fifteen employees, complete with a full-blown bureaucracy making researchers' lives miserable yet making absolutely no difference to ethical progress in animal use.

What are the ethical issues emerging from animal research? Both Plato and Hegel have argued that at least part of a moral philosopher's job is to help draw out and articulate nascent and inchoate thought patterns in individuals and society. In keeping with this notion, several philosophers, beginning in the 1970s, made explicit a number of moral reservations about human uses of animals in general, including invasive animal use in research and testing, and thereby helped draw out the moral queasiness at such use that had gradually developed in society. This task was first engaged by Peter Singer in 1975 as a chapter in his *Animal Liberation,* wherein he challenged the moral justification for a great deal of animal use, including the moral permissibility of harming animals to advance scientific knowledge. Singer's discussion of research on animals articulated widespread social reservations about such *use* of animals, and the book is still in print. In 1982, my *Animal Rights and Human Morality* again challenged the morality of hurting animals in research and also pointed out the inadequacy of the *care and husbandry* provided to such animals, leading to additional suffering that was not only not part of the research, but also, in many cases, inimical to its purposes. Additional work by philosophers Tom Regan, Steve Sapontzis, Evelyn Pluhar, and numerous others has continued to give prominence to the moral questions of research on animals, aided by a number of scientists such as Jane Goodall, who have come to see the moral issues with clarity.

Although different philosophers have approached the issue from different philosophical traditions and viewpoints, it is possible to find a common thread in their arguments questioning the moral acceptability of invasive animal use. Drawing succor from society's tendency during the past fifty years to question the exclusion of disenfranchised humans such as women and minorities from the scope of moral concern, and the correlative lack of full protection of their interests, these philosophers applied a similar logic to the treatment of animals.

In the first place, there appears to be no morally relevant difference between humans and at least vertebrate animals that allows us to include all humans within the full scope of moral concern and yet deny such moral status to the animals. A morally relevant difference between two beings is a difference that rationally justifies treating them differently in some way that bears moral weight. If two students have the same grades on exams and papers, and have identical attendance and class participation, the teacher is morally obliged to give them the same final grade. That one is blue-eyed and the other is brown-eyed may be a difference between them, but it is not morally relevant to grading them differently.

Philosophers have shown that the standard reasons offered to exclude animals from the moral circle, and to justify not assessing our treatment of them by the same moral categories and machinery we use for assessing the treatment of humans, do not meet the test of moral relevance. Such historically sanctified reasons as "Animals lack a soul," "Animals do not reason," "Humans are more powerful than animals," "Animals do not have language," and "God said we could do as we wish to animals" have been demonstrated to provide no rational basis for failing to reckon animal interests in our moral deliberations. For one thing, while several of the above statements may mark differences between humans and animals, they do not mark *morally relevant differences* that justify harming animals when we would not similarly harm people. For example, if we justify harming animals on the grounds that we are more powerful than they are, we are essentially affirming that "might makes right," a principle that morality is in large measure created to overcome. By the same token, if we are permitted to harm animals for our benefit because they lack reason, there are no grounds for not extending the same logic to nonrational humans, as we shall shortly see. And while animals may not have the same interests as people, it is evident to common sense that they certainly do have interests, the fulfillment and thwarting of which matter to them.

The interests of animals that are violated by research are patent. Invasive research such as surgical research, toxicological research, and disease research certainly harms the animals and causes pain and suffering. But even noninvasive research on captive animals leads to pain, suffering, and deprivation arising out of the manner in which research animals are kept. Social animals are often kept in isolation; burrowing animals are kept in stainless steel or polycarbonate cages; and in general animals' normal repertoire of powers and coping abilities—what I have called their teloi, or natures—are thwarted. Indeed, Tom Wolfle, a leading laboratory-animal veterinarian and animal behaviorist formerly at the NIH, has persuasively argued that animals used in research probably suffer more from the ways in which they are kept for research than from the invasive manipulation they are exposed to within research.

147

The common moral machinery that society has developed for adjudicating and assessing our treatment of people would not allow people to be used in invasive research without their informed consent, even if great benefit were to accrue to the remainder of society from such use. This is the case even if the people to be used were cognitively disabled—infants, the insane, the senile, the mentally retarded, the comatose, and so on. A grasp of this component of our ethic has led many philosophers to argue that one should not subject an animal to any experimental protocol that society would not be morally prepared to accept if performed on a mentally retarded or otherwise intellectually disabled human. There appears in fact to be no morally relevant difference between intellectually disabled humans and many animals—in both cases, what we do to the beings in question matters to them, as they are capable of pain, suffering, and distress. Indeed, a normal, conscious, adult nonhuman mammal would seem to have a far greater range of interests than a comatose or severely mentally retarded human, or even than a human infant.

While we do indeed perform some research on the cognitively disabled, we do not do so without as far as possible garnering their consent or, if they are incapable of giving consent, obtaining such consent from guardians specifically mandated with protecting their basic interests. This was not always the case—witness the infamous Tuskegee syphilis experiments on poor, uneducated, rural African American men over the forty-year time span from 1932 to 1972. They were told only that the government was giving them "free health care," but they were never treated with penicillin, even when it had been proven to cure syphilis. Numerous invasive studies on uninformed humans were performed throughout the twentieth century, until the federal government responded to the Tuskegee revelations by mandating strict rules for research on humans.

Applying such a policy to animals would forestall the vast majority of current research on captive animals, even if the bulk of such research is noninvasive, given the considerations detailed above concerning the violations of animals' basic interests as a consequence of how we keep them. Philosopher Steve Sapontzis has further pointed out that we do have a method for determining whether an animal is consenting to a piece of research: open its cage! (Note that an animal's failure to leave the cage would not necessarily assure consent, however; it might merely demonstrate that a condition like learned helplessness has been induced in the animal.)

The above argument, extrapolated from ordinary moral consciousness, applies even more strongly to the case of animals used in psychological research, where one is using animals as a model to study noxious psychological or psychophysical states that appear in humans—pain, fear, anxiety, addiction, aggression, and so forth. Here one can generate what has been called the psychologist's dilemma: If the relevant state being produced in the animal is analogous to the same state in humans, why are we morally entitled to produce that state in animals when we would not be so entitled to produce it in humans? And if the animal state is not analogous to the human state, why create it in the animal at all?

In short, what entitles humans to use animals in ways that harm, hurt, kill, or distress them in research for human benefit? The logic of our societal ethic for humans does not allow humans to be used in such a way. We cannot use prisoners, cognitively disabled persons, unwanted children, dangerous psychopaths, or other socially disvalued human beings in invasive ways without informed consent for the benefit of the majority or of society as a whole. The researchers responsible for the Tuskegee experiments studying syphilis in black male prisoners without informed consent argued that such people were "worth less" than "normal" citizens and thus their interests could be sacrificed for the good of the majority. It is well known that these arguments were categorically rejected when the nature of the study was revealed during the 1970s and in fact prompted detailed federal restrictions on the use of human subjects in research.

I noted earlier that as a utilitarian, Peter Singer in his moral critique of animal research focused on pleasure and pain. As I also indicated, such a critique is far too simplistic. Although no one is quite sure how many animals are utilized in invasive research, since rats, mice, and birds are not included in federal inventories of animal research, the research community itself has estimated that only about 10 percent of animals so used suffer pain. However, as we have stressed throughout this book, animal suffering cannot be limited to physical pain, measurable as utilitarians do, on one axis. Research animals suffer if they are not kept under innumerable conditions determined by their teloi. Nocturnal animals are often kept under twenty-four-hour light cycles; ways in which animals have evolved to nourish themselves are rarely if ever respected under laboratory conditions; social animals are often kept in isolation; spatial boundaries within which animals normally roam are never respected in laboratories; normal weaning of infants from mothers is virtually never respected; and so on. Indeed, normal freedom of movement is by definition not respected in captivity. As we mentioned earlier, animals like coyotes will chew their legs off when they are trapped to regain freedom of movement, which, by virtue of this example, is more important to them than physical pain.

It is probably for the set of reasons detailed above that there are fewer works defending the use of animals in research than criticizing it. One book that did attempt to provide a systematic justification for animal use in research, *The Case for Animal Experimentation*, by Michael A. Fox (1986), was repudiated by its author within months of publication. Nonetheless, there are certain arguments that are frequently deployed by defenders of animal research. Primary among them are the argument from benefits, the argument that moral concerns of the sort required to question animal research apply only to humans, and the argument from experimenting on cognitively disabled humans.

The first argument in defense of research on animals that we shall consider, the argument from benefits, is as follows: Research on animals has been intimately connected with new understanding of disease, new drugs, and new operative procedures, all of which have produced significant benefits for humans and for animals. These significant results and their attendant benefits would have been unobtainable without animal use. Therefore animal research is justified.

Critics of animal research might (and do) attack this argument in two ways. First of all, one might question the link between premises and conclusion. Even if significant benefits have been garnered from invasive animal use, and even if these benefits could not have been achieved in other ways, it does not follow that such use is justified. Suppose that Nazi research on unwilling humans had produced considerable benefits, as, for example, as some have argued, it did in the areas of hypothermia and high-altitude medicine. It does not follow that we would consider such use of human subjects morally justifiable. In fact, of course, we do not. Indeed, there are significant numbers of people in the research community who now argue that the data from such experiments should never be used or even cited, *regardless of how much benefit flows from its use.* The only way for defenders of animal research to defeat this counterargument is to find a morally relevant difference between humans and animals that stops our extending our social-consensus ethic's moral concern for human individuals to animals.

Second, one can attack the argument from benefits in its second premise, namely, that the benefits in question could not have been achieved in other ways. This is extremely difficult to prove one way or the other, for the same reasons that it is difficult to conjecture what the world would have been like if the Nazis had won World War II. We do know that as social concern regarding the morality of animal research mounts, other ways *are* being found to achieve many of the ends listed in our discussion of the uses of animals in research.

The only plausible sort of argument in defense of invasive use of human beings is the utilitarian one, which posits that such use generates more benefit than cost, a claim that society has categorically rejected regarding research on humans. But perhaps, in the case of animals, such an argument is socially acceptable. If that is the case, we are led to another level of ethical concern about the use of animals in scientific experimentation. If the only justification for such use is the benefit it provides, and only when it far outweighs the cost to the animals, then it follows that the only allowable animal use in experimentation would be that that clearly and demonstrably provides greater benefit to humans than the cost to the animals. And this is clearly not the current state of affairs. Animals are deployed in painful ways in myriad experiments that do not provide significant benefit. These experiments range from toxicological experiments that provide only some legal protection for corporations from lawsuits regarding product liability, to experiments in pursuit of new weaponry, to inflicting learned helplessness on animals allegedly to model human depression (illegal in Great Britain), to seeing how many bites an "intruder" animal into an established animal colony sustains, to numerous other experiments augmenting knowledge that appears to be of no practical value.

The second argument commonly used to defend animal research is the argument that moral concerns of the sort required to question animal research apply only to humans. This approach is, in essence, an attempt to provide what I indicated was necessary to buttress the argument from benefits. Philosopher Carl Cohen made such an attempt in a 1986 *New England Journal of Medicine* article, "The Case for the Use of Animals in Biomedical Research," which is generally considered by the research community to be the best articulation of their position.

One of Cohen's chief arguments can be reconstructed as follows (the argument is specifically directed against those who would base condemnation of animal research on the claim that animals have rights, but can be viewed as applying to our earlier discussion of the general argument against invasive animal use): Only beings who have rights can be said to have sufficient moral status to be protected from invasive use in research. Animals cannot reason, respond to moral claims, and so forth—necessary conditions for being rights-bearers. Therefore they do not have rights and so cannot be said to be morally protected from invasive use.

The problems with this argument are multiple. In the first place, even if the concept of having a right (or of having sufficient moral status to protect one from being used cavalierly for others' benefit) arises only among rational beings, it does not follow that its use is limited to such beings. Consider an analogy. Chess may have been invented solely for the purpose of being played by Persian royalty. But given that the rules have a life of their own, anyone can play it, regardless of the intention of those who created the rules. Similarly, rights may have arisen in a circle of rational beings. But it does not follow that such rational beings cannot reasonably extend the concept to beings with other morally relevant features. In fact, that is precisely what has occurred in the extension of rights to nonrational humans.

To this, Cohen replies that such extension is legitimate, while extension to animals is not, since nonrational humans belong to a kind that is rational. The obvious response to this, however, is that, by Cohen's own argument, it is being rational that is relevant, not belonging to a certain kind. Further, if his argument is viable, and one can cavalierly ignore what is by hypothesis the morally relevant feature, one can turn it around on Cohen. One could argue in the same vein that since humans are animals, albeit rational ones, and other animals are animals, albeit nonrational ones, we can ignore rationality merely because both humans and animals belong to the same kind (that is, animal). In short, his making an exception for nonrational humans fails the test of moral relevance and makes the arbitrary inclusion of animals as rights-bearers as reasonable as the arbitrary inclusion of nonrational humans.

A detailed exposition of and response to such a strategy is impossible to undertake here (the philosopher Evelyn Pluhar has engaged this task in *Beyond Prejudice* [1995]). However, the following points can be sketched: Moral status (and therefore rights) in our social ethic for people is based on what matter greatly to people. Certainly pain matters to animals as to people, making it a sufficient condition for moral status. First, a heavy burden of proof exists for those who would convince common sense and common morality that animals cannot feel pain. Even the anticruelty ethic took animal pain for granted. Second, animals' not being able to feel pain while humans can would make the appearance of pain and other modes of awareness in humans an evolutionary miracle. Third, the neurophysiological, neurochemical, and behavior evidence militates in favor of numerous similar morally relevant mental states such as pain in humans *and* animals. Fourth, if animals are truly devoid of awareness, much scientific research would be vitiated, for example, pain research conducted on animals and extrapolated to people.

One possible way to exclude animals from direct moral status, and thereby justify invasive research on them, is a philosophically sophisticated exposition of the claim we discussed by Cohen that morality applies only to rational beings. This position, which has its modern roots in Hobbes but in fact was articulated even in antiquity, was more recently thrust into prominence by the work of Harvard philosopher John Rawls. It has been directly applied to the question of animals' moral status by British philosopher Peter Carruthers, who has also advanced a neo-Cartesian argument against animal mind in his book *The Animals Issue* (1992). Interestingly enough, Carruthers's contractual argument is independent of his denial of consciousness to animals. Even if animals are conscious and feel pain, Carruthers believes that the contractual basis for morality excludes animals from the moral status necessary to question the moral legitimacy of experimentation on them.

According to Carruthers, morality is a set of rules derived from what rational beings would rationally choose to govern their interactions with one another in a social environment, if given a chance to do so. Only rational beings can be governed by such rules and adjust their behaviors toward one another according to them. Thus, only rational beings, of which humans are the only example, can "play the game of morality," so only they are protected by morality. Animals thus fall outside the scope of moral concern. The only reasons for worrying about animal treatment are contingent ones, namely, that some people care about what happens to animals, or that bad treatment of animals leads to bad treatment of people (as Thomas Aquinas argued), but nothing about animals in themselves is worthy of moral status. Further, the above contingent reasons for concern about animal suffering do not weigh heavily enough to eliminate research on animals.

There are a variety of responses to Carruthers. In the first place, even if one concedes the notion that morality arises by hypothetical contract among rational beings, it is by no means clear that the only choices of rules such beings would make would be to cover only rational beings. They might also decide that any rules should cover any beings capable of having negative or positive experiences, whether or not they are rational. Second, even if rational beings intend the rules to cover only rational beings, it does not follow that the rules do not have a logic and life of their own that lead to adding other beings to the circle of moral concern, as indeed seems to be happening in social morality today. Third, Carruthers seems to assume that according moral status to animals requires that the status be equal to that of humans, "yet," he says, "we find it intuitively abhorrent that the lives and suffering of animals should be weighed against the lives or suffering of human beings" (1992, 195). But it is not at all clear that contractualism, even if true, could not accord animals sufficient moral status to prohibit experimenting on them, yet not say they were of equal moral value to humans. Further, as Steve Sapontzis (1987) has pointed out, Carruthers's argument is circular. He justifies such uses as research on animals by appeal to contractualism, and he justifies contractualism on the grounds that it renders morally permissible such uses as research on animals.

The final defense of research on animals that we shall consider is the utilitarian one advanced by the philosopher R. G. Frey (1980), the argument from experimenting on cognitively disabled humans. Unlike the previous arguments, it is a tentative one, offered up in a spirit of uneasiness. Frey's argument essentially rests upon standing the argument from cognitively disabled humans on its head. Recall that this argument says that animals are analogous to such humans as the mentally incapacitated, the comatose, the senile, the insane, and so on. Since we find experimenting on such humans morally repugnant, we should find experimentation on animals equally repugnant.

Frey's argument reaffirms the analogy but points out that, in actual fact, many physically and intellectually healthy animals have richer and more complex lives, and thus have *higher quality* lives, than many cognitively disabled humans do. The logic of justifying research on animals for human benefit (which assumes that humans have more complex lives than animals, and thus more valuable lives) would surely justify doing such research on cognitively disabled humans who both have lower qualities of life than some animals do and are more similar physiologically to healthy humans and thus are better research "models." If we are willing to perform such research on cognitively disabled humans, we are closer to justifying similar research on animals.

Obviously, the force of Frey's argument as a defense of research depends upon our willingness relentlessly to pursue the logic by which we (implicitly) justify animal research and to apply the same justification to using humans not different from those animals in any morally relevant way. As Frey himself affirms, there are some "contingent" (i.e., not logically necessary) effects of deciding to do research on cognitively disabled humans as well as on animals that would work against such a decision. He cites the emotional (rather than rationally based) uproar and outrage that would arise (because people have not worked through the logic of the issue) and presumably such other responses as the knee-jerk fear of a slippery slope leading to research on other humans. But, in the end, such psychological rather than moral/logical revulsion could conceivably be overcome by education in and explanation of the underlying moral logic.

Arguably Frey's argument fails as a defense of research on humans and ends up serving those who originally adduced the argument from cognitively disabled humans as a proof *against* research on animals. If people do see clearly and truly believe that doing research on animals is (theology aside) exactly morally analogous to doing research on cognitively disabled humans, they are, in our current state of moral evolution, likelier to question the former than to accept the latter.

In fact, Frey's argument very likely serves to awaken a primordial component in human moral psychology, revulsion at exploitation of the innocent and the helpless—animals and cognitively disabled humans being paradigm cases of both. In a society that increasingly and self-consciously attempts to overcome such exploitation, experimentation on cognitively disabled humans, in fact often practiced in the past, along with experimentation on powerless humans, is not a living option. In sum, then, the force of Frey's argument is not to justify research on animals but rather to underscore its morally problematic dimension.

Thus, the only argument in defense of animal research that seems at all cogent is the argument from benefits discussed above. A utilitarian thinker might argue that with regard to animal subjects or human subjects utilized in research, even invasive research, such research is justified if the benefits to sentient beings, humans or animals, outweighs the cost to the subjects. Peter Singer, for example, a consistent utilitarian, has argued that certain invasive neurological research on nonhuman primates is justified because of the large numbers of humans whose health has been markedly improved as a result of that research.

Our societal ethic, embedded in our laws, does not of course accept such an argument regarding research on humans, and checks a purely utilitarian ethic by use of the deontological notion of rights, protecting individual humans from having their basic interests infringed upon even for the sake of the general welfare. Hence, as social consciousness of the ethics of research on humans evolved in the wake of revelations of human abuse in research both in the United States (such as the Tuskegee syphilis experiments) and in Nazi Germany, society roundly condemned Nazi research that was scientifically and medically valuable, such as hypothermia and high-altitude medical research, along with the patently useless research performed by Josef Mengele.

For the sake of argument, let us assume that invasive animal research is justified only by the benefit produced. It would then seem to follow that the only morally justifiable research would be research that benefits humans (and/or animals). But there is in fact a vast amount of research that does not demonstrably benefit humans or animals. Much of the behavioral research conducted, weapons research, and toxicity testing as a legal requirement are obvious examples, as is much basic research that is invasive but has no clear benefit. Obviously a certain amount of research meets that test, but a great deal does not. Someone might respond that "we never know what benefits might emerge in the future" and appeal to serendipity or unknowns. But if that were a legitimate point, we could not discriminate in funding between research likely to produce benefit and that which is unlikely to produce benefit; yet we do. If we appeal to unknown but possible benefits, we are no longer funding based on *known benefit*. We do in fact weigh cost versus benefit in human research and in animal research. Why not also weigh cost to the animal subject as a relevant parameter?

Thus we find a second major moral issue in animal research. To recapitulate: The first issue arises from the suggestion that any invasive research on an object of moral concern is morally problematic. In response, researchers invoke the benefits of research. Even assuming this is a good argument, it gives rise to another moral issue: *Why do we not do only that animal research that clearly produces more benefit than cost to the animals?* It is clear that this is not the case. It is well known that most publications of animal research results are almost never cited in later research, thereby casting doubt on the value of much of the research being done. As Andrew Knight (2012) has deftly demonstrated, this is even true of the most highly valued area of animal research, research on chimpanzees, allegedly the closest animal relation of humans. Also highly relevant to this point is the fact that researchers overlook the costs to the animals they use that results from their appalling lack of knowledge of animal metabolism, physiology, and behavior. If one never quantifies the cost to animals, one cannot decide whether the benefit exceeds it. In particular, most researchers are *significantly ignorant of animal telos and the myriad considerations flowing from it that deform and invalidate their results.*

There are vast numbers of aspects of animal telos that most people do not notice. Yet the violations of these interests have major consequences not only for the animals, but for their functioning as models in science as well. Virtually any aspect of animal nature can, if violated, have myriad consequences for the animal's metabolism, physiology, and normal functioning, as well as for the animal's welfare. There is a vast literature detailing such considerations, as we will shortly indicate. Yet the average researcher knows nothing of this.

In the mid-1980s, I was approached by an editor of an anthology on the ethics of animal research. Knowing that I had a fair amount of knowledge concerning research, he asked me to discuss the most outrageous situation regarding animals used in science. I unhesitatingly explained that one could get an MD/PhD degree in an area using laboratory animals and never learn anything about the animals one uses except that they model a particular disease or syndrome, and get funded to establish a research program while knowing literally nothing of the biological and psychological needs of the animals in question. This explains the extraordinary anecdote I heard on three different continents from veterinarians whose expertise was laboratory animals. In each case, the veterinarians had been approached by researchers complaining that they had been supplied with "sick dogs" by the veterinarian. The basis for these complaints was the fact that the animals in question all had "fevers" above 98.6 degrees Fahrenheit. What the researchers did not know, of course, is that *the normal body temperature of the dog is 101.5!*

Laboratory animals were kept in accordance with human convenience, not in accordance with their biological natures, or their teloi, which in most cases caretakers were not even aware of. Cages were designed for ease of cleaning, not for fitting animal needs. An excellent example is provided by rats and mice, who by nature are nocturnal, burrowing creatures. Yet, for human convenience, researchers kept them in lighted rooms twenty-four hours per day. So inimical was this to rat nature that they experienced retinal deterioration under these conditions. Researchers kept monkeys, who are highly social, in isolation in cages totally devoid of stimulation. Not only were animal natures not respected, they were not even known, and thus no attempts were made to accommodate them. So impoverished were environments for primates that it took a 1986 federal law to demand housing for nonhuman primates that "enhances their psychological well-being."

This sort of neglect was true for virtually every species used in research. Dogs were often caged singly with no attention paid to their socialization with humans. Cats, who are creatures that jump and climb, were not provided with perches. Whether in agriculture or in research, absolutely no attempts were made to respect animals' natures, thereby putting their welfare in a highly compromised situation. In animals used in research, absolutely no attention was paid to the stressfulness of environments, despite the fact that even tiny variations in environment wreak havoc with physiological and metabolic variables. Increasing the cage size for rodents has been shown to reduce the toxicity of amphetamine by 50 percent. The LD50 test for acute toxicity of amphetamine in rats, that is, the amount of amphetamine needed to provide a lethal dose to 50 percent of them, increases by a factor of seven when the rats are caged in groups of twelve, compared to when they are caged singly. Virtually monthly, examples are published illustrating the profound effect of housing conditions. For example, in 2014, a study in *Nature Methods* showed that the stress of rats is considerably elevated when their human caretakers are male rather than female. Another 2014 study, in *BMC Medicine,* demonstrated that stresses, such as merely restraining pregnant rats for as little as twenty minutes, can have measurable physiological and behavioral effects on *granddaughters of those rats.* Thousands of examples of these sorts of effects have been published since researchers began paying more attention to housing and care conditions, even showing reductions in certain diseases when rabbits are socially housed.

It is, then, evident that ignoring animal telos in animal housing has profound effects upon one's research results. Thus, the requirements of good science dictate that attention be paid to animals' natures lest important variables relevant to the object of study be ignored or distorted. Remarkably, compliance with emerging societal ethics for animal treatment creates a demand for precisely the same attention to animal nature. As I mentioned earlier, it is rare that the requirements of science and societal ethics converge. To minimize the stress that results from conflict between the requirements of research and animals' needs and natures, we should come as close as possible to creating conditions for the animals that meet the dictates of their teloi.

Astonishing as it may be that researchers ignored both ethical and scientific reasons for respecting telos, the knowledge that the failure to respect animal nature wreaks havoc with scientifically relevant variables is not new. Despite the historical failure to utilize analgesia in painful experiments, as discussed earlier, simple common sense should have suggested to researchers that an animal suffering pain is not a biologically normal animal. Metabolic variables and physiologic variables are disturbed by the failure to control pain, as are immunological parameters and, correlatively, resistance to disease. There is also data showing the converse. Animals who are treated exceptionally well have a significantly higher degree of reproductive success, and rabbits who are treated with exceptional care develop significantly fewer atherosclerotic lesions when fed a high cholesterol diet. Incredibly, the attempt to develop a cage for mice that was considerably more congenial to mouse telos beginning in the 1990s was met with extreme criticism on the grounds that it would vitiate, invalidate, or at least call into question all the previous mouse data accumulated from mice housed in cages designed for human convenience, not mouse nature.

One of the best and most careful papers documenting the degree to which variables not considered by animal researchers could wreak havoc with data is a 1982 British article by G. Clough, "Environmental Effects on Animals Used in Biomedical Research." In this superb paper, the author examines in some detail the degree to which animals' responses are "influenced by far more environmental conditions, and often to a far greater degree, than is appreciated by very many investigators, perhaps even the majority" (487). Clough also remarks that an "alarmingly low percentage of authors mention such basic information as ambient temperature and lighting regimes" (488). He discusses the following environmental variables and illustrates their effects on animals under varying conditions: temperature, relative humidity, air movement, air quality (including the physical state of the air and the content of the air), and light (including intensity, wavelength, and photoperiod).

There are of course many variables that the author fails to mention, such as personality of the technician, crowding, ambient pheromones, and so on.

Even something as putatively trivial as uncorking a bottle of ether in a laboratory can have measurable effects on animals. The key point to reiterate is that the failure to create environments accommodating research animals' teloi is as harmful to science as to ethics. Thus we find in the area of animal research a scientific as well as an ethical reason to learn as much about animal needs and natures as possible to assure the validity of the animal as a model. As is well known, the more we minimize and control extraneous effects on animals, the sounder our science will be. Thus by accommodating the dictates of an animal's telos, we do good while we do well. We find we have sound reasons from both viewpoints to unify ethical and scientific considerations. And once again, we find a powerful reason even from the scientific ideology standpoint for viewing animals in terms of their teloi, their biological and psychological natures. If we look at animals in a reductionistic way, strictly in terms of physics and chemistry, we not only lose sight of what makes an animal what it is, we also guarantee deformation of the variables central and vital to our inquiry.

Attendant on the points we have just made, a startling conclusion emerges for animal research. As indicated earlier, the laws governing animal research are implemented by animal care and use committees consisting overwhelmingly of scientists who are suffused with scientific ideology. In other writings, I have stressed that this system represents an extreme case of the fox watching the henhouse. This is not to suggest that we need to distrust scientists on committees. In my experience, there is a strong tendency on the part of such scientists to abide by federal law. It is rather that scientists tend to see the world through ideological glasses based on the way in which they are both trained and suffused. Thus they will inevitably miss relevant considerations filtered out by that ideology. In addition, scientific research is largely done with public money in the name of public interest. Yet when ordinary people see the world through commonsense, ethics-based lenses, they see a good deal that scientists do not and about which scientists tend not to care. As I have indicated throughout this book, ordinary people now care a great deal about the metaphysics of ordinary common sense and about the ethical issues to which science gives rise. For these reasons, ordinary people should have a significant voice in deciding what research is to be funded and how it is to be conducted.

To the standard complaint that only scientists can understand what the scientists do, there is a simple rejoinder. Science has become so specialized that most scientists are essentially laypeople outside their own immediate area.

In addition, the purpose of a piece of scientific research can be explained to intelligent laypeople by intelligent scientists. Only in advanced and esoteric areas of physics can this not be done, and those areas do not use animals. If public money is the fuel that drives science, in a democratic society the public should have a significant voice in how that money is expended, in terms of both what areas are researched and how ethically that research is conducted.

There are some limited but promising indicators that the research community is moving in that direction. In the 2013 edition of the NIH *Guide to the Care and Use of Experimental Animals,* there is a directive that social animals should be housed in a manner that respects their sociality. There is also a directive promoting the adoption of adoptable animals at the ends of research projects. Furthermore—and this is a highly significant tip of the hat to common sense and social morality—the cost to the animals utilized in research projects should be weighed against the benefit likely to be produced by the research. As we approach a new era in society wherein the general public concerns itself far more dramatically with both the logic and ethics of research, and no longer fully abides by "Trust me, I'm a scientist," the need for incorporating ethics into science becomes far more pressing.

The realization that animals utilized in research are not simply inert tools but rather participants in research who are objects of moral concern must occasion far more concern with the ethical nuances of animal research, as well as with other ethical dimensions of science. For example, it is likely that the public will be far less accepting of weapons research performed on animals. Until very recently, nonhuman primates, even chimps, were the subjects of choice for such research. With the scientific community and the public all over the world seriously questioning the right of researchers to use nonhuman primates for such purposes, questioning will be greatly accelerated, including numerous attempts around the world to provide animals with a higher moral status. (In 2015, the NIH announced that it would no longer use or keep chimpanzees.) A similar point is also being made internationally regarding the use of dogs in research. To be sure, there are many animals also logically worthy of higher moral consideration who do not enjoy the privileged status of primates and dogs, but there is evidence that social thought is moving in the right direction.

In sum, there is reason to believe that animal research may acquire an enhanced moral patina more quickly than confinement agriculture does. This is a good thing, for it is very likely to constitute a tide that floats all boats.

Genetic Engineering and Telos

ANY DISCUSSION OF animal ethics in relation to telos automatically gives rise to the issues occasioned by genetic engineering of animals for production purposes or for research. The question that arises is whether it is wrong intrinsically to genetically modify animal telos. In the traditional, Aristotelian, account of animal telos, the question would not arise since for Aristotle teloi are fixed and immutable natural kinds. For Aristotle there is no modification of telos possible, because if one could effect such modification, knowledge would be impossible. For Aristotle, the universe is inherently knowable. It is not that Aristotle had no acquaintance with thinkers who postulated modifiability of natural kinds. Empedocles, for example, expressed an embryonic theory of evolution by natural selection. In Empedocles's cyclical cosmogony, the material components of the universe combine and disintegrate in accordance with the actions of two principles he called "love" and "strife," which are forces of attraction and repulsion. This occurs in a regular and cyclical way. For example, random combinations of basic elements produce teratological entities such as a man with the head of an ox (a Minotaur). Being unsuited for survival, such organisms are selected out. These epochs have been recorded in myths and legends.

Aristotle rejected such evolutionary accounts on the grounds that we see no evidence that these things occur. It has been suggested that Aristotle, as a naturalist, would probably have encountered such regularly occurring entities as fossilized fish. But, in accordance with his powerful theoretical bias, he would have seen such things not as fossils, remnants of things past, but rather as stone fish, themselves a different natural kind.

Nonetheless, in today's scientific worldview, natural kinds are not fixed and immutable but stages along a gradual series of mutations, occurring very slowly but nonetheless leading to variations in natural kinds significant enough to count as new species—witness the modern account of the evolution of the horse.

If minute changes in genetic traits favorable to survival lead to species change, there is no good reason to suggest that human intervention, intentional or unintentional, cannot be a vector in the evolution of species. We have in fact drastically altered species in an intentional way ever since we have been able to do so. We have created new plant species in abundance through hybridization—the tangelo and numerous subtypes of orchid are examples of such genetic manipulation. Indeed, it is estimated that 70 percent of grasses and 40 percent of flowering plants represent new species created by humans through hybridization, cultivation, preferential propagations, and other means of artificial selection. (We have as yet to produce a new species of animal, but the barriers to doing so are technological in nature, soon to be overcome.)

One of the first and most pronounced ethical objections to genetic engineering has been the largely theologically based claim that genetic engineering violates "species integrity" and is thus "intrinsically wrong." This is of course a variation on the old dictum that was a mainstay of horror movies: "There are certain things humans were not meant to do." Such a claim perhaps makes sense in the metaphysics that proclaims that God created the universe out of immutable natural kinds, but not in the metaphysical worldview of science and evolution. In fact, numerous theologians of all denominations have had the chutzpah to proclaim that genetic engineering (or cloning, in reference to Dolly, the first cloned sheep) "violates God's will."

In 1995 I published the first book on the ethical issues attendant upon genetic engineering of animals, *The Frankenstein Syndrome*. I discovered rather quickly that there was no place to turn to get an idea of the scope of these issues. Since the scientific community was captive to the ideological belief of "value-free science," no one in that community was addressing these issues. The general public, in contradistinction to the research community, was extremely concerned about the ethics of biotechnology but could hardly have been more scientifically illiterate and was totally suspicious of science and particularly of anything "messing with nature," which I called "the Frankenstein syndrome" in my book.

We have already discussed the US public's scientific illiteracy. There is little reason to believe things have improved in recent years. The combination of scientific ideology, or the common sense of science, on the part of scientists and public illiteracy and suspicion concerning science created a perfect storm, resulting in what I have called a "Gresham's law for ethics." Gresham's law, it will be recalled, was first articulated by Renaissance economist Thomas Gresham, who affirmed in essence that "bad money drives good money out of circulation." Consider the post–World War I German economy. The Deutschmark was so devalued by inflation that it took a wheelbarrow full of them to purchase a loaf of bread. Thus if one owed ten thousand marks on a piece of real estate, one would not pay that debt in gold. Rather, people would pay with valueless paper. Similarly, I have argued that "bad ethics drives good ethics out of circulation." Bad ethics, such as the claim that genetic engineering violates God's will, is far more seductive than are legitimate concerns about the welfare of a genetically engineered animal, and thus it tends to seize center stage. And once a piece of bad ethics has established itself, it is extremely difficult to dislodge. Claims like this enjoy the appeal of the exotic.

There are three kinds of allegedly ethical claims about genetic engineering that are worthy of consideration. The first is some variation on "intrinsic wrongness" of biotechnology, such as that it violates God's will. Such claims are invariably without a basis. The second sort of issue raises the question of societal or biological risks and dangers brought into play by the technology. Strictly speaking, of course, this is less an ethical issue than a prudential one. No one benefits if the technology occasions harm or disaster. The most genuine ethical issue, the third kind, concerns what I have termed "the plight of the creature" and involves harms that may befall a genetically engineered animal. If the animal is genetically altered in such a way as to impair its welfare, that is a legitimately ethical issue, even if human benefits are correlatively produced. This is particularly true in a time when society has begun to take issues pertaining to animal welfare much more seriously than at any other time in our history. However, such legitimate questions tend to become eclipsed by bad ethics.

Through the vehicle of transgenic animals, one can, for example, study the functions of various genes in organisms by ablating genes from the genome of an animal that is relatively familiar. Or, conversely, one can add new genes taken from radically diverse organisms. The effects of such manipulations are unpredictable and may result in serious welfare problems for the experimental organisms.

For many people, the genome encoding the telos of a certain kind of animal is sacred and inviolable. This is based in a radical misunderstanding. Given an animal possessed of a certain telos, one should not violate the interests flowing from that telos. But I would argue that it is not wrong to change the genes determining an animal's nature. For example, if a certain sort of animal often suffers from a genetic disease, I see no problem with replacing the gene in question to change the kind of life the animal leads in a positive way. The problematic issue therefore lies not in changing telos per se. Some changes can be very positive, while others would seriously harm the animal's quality of life. The problem, rather, lies in violating what I call the Principle of Conservation of Welfare. This principle asserts that if one is genetically modifying a given telos, *the resulting animal should be no worse off after genetic modification than the parent stock would have been without the genetic modification, and ideally the animal should be better off.* (It is interesting and gratifying that when I enunciated this principle at a USDA conference on genetic engineering of animals, the audience response was almost unanimously positive.) The moral problem of genetically modifying an animal (or more accurately, a kind of animal) arises only if the animal's welfare is compromised or diminished by the genetic modification, in which case it should not be done.

What does genetic engineering portend for animal welfare? Given our history of animal use, the prospects are not positive. After all, after some ten thousand years of animal agriculture based on honoring the ancient contract and deeply respecting and relying on animal husbandry, it is remarkable how quickly the rise of industrial agriculture could subvert these deeply ingrained principles for the sake of profit and productivity. Efficiency, productivity, and profit quickly came to dominate animal agriculture, with good husbandry rapidly relegated to, at best, some antiquated nostalgia and, at worst, something to be overcome.

Biotechnology, in the same vein, provides us with virtually complete control over animals. For example, we were historically constrained in our ability to study human genetic diseases in animal models by the natural and adventitious occurrence of the relevant mutations. Now we can simply modify relevant genes at will. With current technology, one can in principle create animal models for all human genetic diseases, no matter how horrendous and no matter how much suffering their creation entails.

To flesh out our discussion with a real example, let us examine the first attempt to produce an animal "model" for human genetic disease by transgenic means, that is, the development, by embryonic stem-cell technology, of a mouse that was to replicate Lesch-Nyhan syndrome, or hypoxanthine-guanine phosphoribosyltransferase (HPRT) deficiency. Lesch-Nyhan syndrome is a particularly horrible genetic disease that leads to a "devastating and untreatable neurologic and behavioral disorder." Patients rarely live beyond their third decade, and suffer from spasticity, mental retardation, and choreoathetosis, which is uncontrollable, jerky, spasmodic writhing. The most unforgettable and striking aspect of the disease, however, is the patient's irresistible compulsion to self-mutilate, usually manifesting itself as biting fingers and lips. The following clinical description by W. N. Kelley and J. B. Wyngaarden conveys the terrible nature of the disease:

> The most striking neurological feature of the Lesch-Nyhan syndrome is compulsive self-destructive behavior. Between 2 and 16 years of age, affected children begin to bite their fingers, lips and buccal mucosa. This compulsion for self-mutilation becomes so extreme that it may be necessary to keep the elbows in extension with splints, or to wrap the hands with gauze or restrain them in some other manner. In several patients mutilation of lips could only be controlled by extraction of teeth.
>
> The compulsive urge to inflict painful wounds appears to grip the patient irresistibly. Often he will be content until one begins to remove an arm splint. At this point a communicative patient will plead that the restraints be left alone. If one continues in freeing the arm, the patient will become extremely agitated and upset. Finally, when completely unrestrained, he will begin to put the fingers into his mouth. An older patient will plead for help, and if one then takes hold of the arm that has previously been freed, the patient will show obvious relief. The apparent urge to bite fingers is often not symmetrical. In many patients it is possible to leave one arm unrestrained without concern, even though freeing the other would result in an immediate attempt at self-mutilation.
>
> These patients also attempt to injure themselves in other ways, by hitting their heads against inanimate objects or by placing their extremities in dangerous places, such as between spokes of

a wheelchair. If the hands are unrestrained, their mutilation becomes the patient's main concern, and effort to inflict injury in
some other manner seems to be sublimated. (Kelley and Wyngaarden, 1983)

At present, "there is no effective therapy for the neurologic complications
of the Lesch-Nyhan syndrome," according to Kelley and Wyngaarden, who
boldly suggest in their chapter on HPRT-deficiency diseases that "the preferred form of therapy for complete HPRT deficiency [Lesch-Nyhan syn-
drome] at the present time is prevention," that is, "therapeutic abortion."
This disease is so dramatic that I predicted in 1976 it would probably be the
first disease for which genetic researchers would attempt to create a model
by genetic engineering. Researchers have, in fact, sought animal models for
this syndrome for decades and have created rats and monkeys who will self-
mutilate when caffeine drugs are administered to them, though they do not
have Lesch-Nyhan. It is thus not surprising that this was the first disease genetically engineered by embryonic stem-cell technology, in mice. But to the
surprise of the researchers, these animals, although they lacked the HPRT
enzyme, were phenotypically normal and displayed none of the metabolic or
neurologic symptoms characteristic of the disease in humans. The reason for
the failure of this transgenic "model" has been suggested to be the presence
of a backup gene for xanthine metabolism in mice, though other research has
cast doubt on this notion. Though an asymptomatic mouse is still a useful
research animal, for example to begin to test gene therapy, clearly a symptomatic animal would, as a matter of logic, represent a higher-fidelity model of human disease, assuming the relevant metabolic pathways have been
replicated. Presumably, too, it is simply a matter of time before researchers
succeed in producing symptomatic animals—I have been told in confidence
of one lab that seems to be close to doing so, albeit in a different species of
animal. One may perhaps need to move up to monkeys to achieve replication of the behavioral aberrations.

The practical moral question that arises, then, is clear: Given that researchers will certainly generate such animals as quickly as they are able to
do so, how can one assure that the animals live lives that are not character-
ized by the same pain and distress they are created to model, especially since
such animals will surely be used for long-term studies of the development of
genetic diseases? Or should such animal creation be forbidden by legislation,

the way we forbid multiple uses of animals in unrelated surgical protocols in the United States or as the British forbid learned-helplessness studies?

A similar point can be made regarding genetic engineering of animals deployed in agriculture. An agricultural community willing to confine veal calves in pens where they cannot move so that the flesh stays devoid of muscle and is thus extremely tender will not cavil at using genetic engineering for profit, regardless of the effect on animal welfare. Actual research has been done to create animals by genetic engineering who will put on edible flesh more rapidly than normal. The attempts that have thus far been made to genetically engineer farm animals have generated serious welfare problems. For example, attempts to increase the growth rate and efficiency of pigs and sheep by insertion of modified genes to control growth, while achieving that result, have engendered significant suffering (Pursel et al., 1989). The desired results were to increase growth rates and weight gain in farm animals, reduce carcass fat, and increase feed efficiency. Although certain of these goals were achieved (in pigs, rate of gain increased by 15 percent, feed efficiency increased by 18 percent, and carcass fat decreased by 80 percent), unanticipated effects, with significantly negative impact on the animals' well-being, also occurred. Life-shortening pathogenic changes in pigs, including kidney and liver problems, were noted in many of the animals. The animals also exhibited a wide variety of diseases and symptoms, including lethargy, lameness, uncoordinated gait, bulging eyes, thickening skin, gastric ulcers, severe synovitis, degenerative joint disease, heart disease of various kinds, nephritis, and pneumonia. Further, their sexual behavior was anomalous; females were anestrous and boars lacked libido. Other problems included tendencies toward diabetes and compromised immune function. Sheep with modified genes fared better than pigs for the first six months but then became unhealthy.

There are certain lessons to be learned from these experiments. In the first place, although similar experiments had been done earlier in mice, mice did not show many of the undesirable side effects. Thus it is difficult to extrapolate in a linear way from species to species when it comes to genetic engineering, even when, on the surface, the same sort of genetic manipulation is being attempted.

Second, it is impossible to effect simple one-to-one correspondence between gene transfer and the appearance of desired phenotypic traits. Genes may have multiple effects, and traits may be under the control of multiple genes. The relevance of this point to welfare is obvious: one should be extremely circumspect in one's engineering until one has a good grasp of the physiological mechanisms affected by a gene or set of genes. A good example of the welfare pitfalls is provided by attempts to genetically engineer mice to produce greater amounts of interleukin 4 to study certain aspects of the immune system (Lewis et al., 1993). This, in fact, surprisingly resulted in these animals' experiencing osteoporosis, a disease resulting in bone fragility, clearly a welfare problem.

Another example is provided by an attempt to produce cattle genetically engineered for double muscling (Gordon Niswender, personal communication). Though a calf was born showing no apparent problems, within a month it was unable to stand up on its own, for reasons that are not yet clear. To the researchers' credit, the calf was immediately euthanized. Yet another bizarre instance of totally unanticipated welfare problems can be found in a situation where leglessness and craniofacial malformations resulted from the insertion of an apparently totally unrelated gene into mice (McNeish et al., 1988).

Thus welfare issues arise in research both on genetically engineered agricultural animals and, more drastically, in potential commercial production. The research-animal issues can best be handled with judicious use of anesthesia, analgesia, and, above all, early end points for euthanasia if there is any suffering. The issues associated with mass production of suffering genetically engineered animals must be dealt with in a different way. For this reason, I have proposed the Principle of Conservation of Welfare, raised earlier.

In summary, I have argued that an animal ethic based on telos does not preclude the possibility of genetically engineering animals in a way that changes telos, because it is not the animal's nature that is sacred and inviolable. It is rather the interests that flow from the animal's nature. It is certainly open to us conceptually to genetically engineer animal teloi in a manner that increases animal welfare, as long as the engineering in question respects the Principle of Conservation of Welfare.

It is not the possibility of genetic engineering in itself therefore that should occasion moral concern. It is rather the new possibilities of creating defective and suffering animals either for purposes of research or for commercial purposes. In the current moral landscape, there is no clear mechanism for preventing the creation of defective animals. Any attempt to prevent such creation should begin with the categorical rejection of bad ethics when discussing genetic engineering—for example, invoking God's will—and instead focus attention on the suffering created, intentionally or unintentionally, in the course of genetically engineering animals. Alas, the public pays little attention to clarification of the issues, and the research community is ethically illiterate.

Conclusion

EMERGING SOCIETAL CONCERN for the billions of sentient creatures with whom we share the earth may well be seen in the future as the major ethical revolution of the twentieth century. As such, it was extremely important to me to engage my fellow citizens in this crusade. When one considers the staggering amount of unnecessary suffering experienced by animals in all the multifarious uses we impose upon them, it is virtually impossible to remain unmoved. For these reasons, based in forty years of developing these ideas and attempting to put them to practical use, I am convinced that one must establish a strong link between commonsense morality and animal ethics. I have attempted to do this by invoking Plato's notion of recollection and Aristotle's notion of telos, both of which, when properly understood, accord extremely well with the thinking of ordinary people, as I have tried to demonstrate in this book. Anyone who has been around animals is aware of their inherent needs and interests, of the significant extent to which these components of their nature or telos matter to them, and of the extent to which the violation of these components causes both physical and psychological suffering to these innocent creatures.

I have exhaustively illustrated in this book some of the major ways through which human use of animals can clash with animal telos; industrialized agriculture and animal research are paradigmatic exemplars. But we must never forget that this need not be the case. For ten thousand years we practiced husbandry-based agriculture that depended on respecting telos. We could do so again, were our perspective not blinded by greed. Indeed, given our cavalier disregard of the environment that sustains us and our total dependence on cheap energy, it is not impossible that we would need at some point to return to husbandry-based, sustainable agriculture, both animal and plant. We have also seen that animal research is most accurate when we respect the animals' needs and natures.

All readers of this book can presumably extrapolate my arguments for respecting telos to more minor animal uses. I am thinking, for example, of such frivolous uses of animals as zoos and circuses. When I was young, clear through my college career, I had no greater joy than going to the zoo, because of my love for animals. But the more reflecting I did on animal ethics, the more it became clear to me that zoos represent a paradigmatic violation of animal teloi. Lions and elephants should not be confined in small cages or pens when their normal range is thousands of acres; killer whales should not be kept in what are in effect backyard swimming pools. So violative are such facilities of animal nature that the animals go mad and become grotesque parodies of their full potential.

On one occasion, I was watching a traveling circus unloading animals to their temporary quarters. One of the animals was a black bear who did nothing but weave compulsively in his cage. I watched him for half an hour and he never did anything but weave, which I understood to be extreme stereotyped behavior. The experience was heartbreaking as I continued to watch. I later found out this was a common animal reaction to an unnatural environment.

On another occasion, I was watching the dolphins prominently displayed in the window tanks at the Mirage Hotel and Casino in Las Vegas. I watched for ten minutes, and then I suddenly began to weep uncontrollably and could not stop, moved by the great gulf between what the animals were doing and what they should have been doing. I suddenly imagined a world in which animals were kept solely for untutored human amusement.

It is necessary for urban humans to learn about animals in order to care about them, but that is not the way. Technological miracles such as video cameras recording animals under their natural conditions can teach us the right things, not merely boasting of our ability to subjugate and control animals while abandoning the priceless gift of empathy.

If, by virtue of considering the arguments in this book, some people are led to take animal suffering more seriously by their own lights, I will have succeeded in my mission, while at the same time, I hope, eliciting in them wonder and delight occasioned by the other lives with whom we share the earth.

References

American Veterinary Medical Association. 1987. Panel Report on the Collo-
quium on Recognition and Alleviation of Animal Pain and Distress.
Journal of the American Veterinary Medical Association 191 (10):
1186-1191.

Aristotle. 1941. *Metaphysics.* In *The Basic Works of Aristotle,* ed. R. M. McK-
eon. New York: Random House.

Bentham, Jeremy. 1996. *An Introduction to the Principles of Morals and Legislation.* Oxford: Clarendon.

Black, Keith. 2004. "Scientific Illiteracy in the U.S." *Cedars-Sinai Neurosciences Report,* Fall.

Braybrooke, David. 1996. "Ideology." In *Encyclopedia of Philosophy,* vol. 2, ed.
Paul Edwards. New York: Simon and Schuster.

Buytendijk, F. J. J. (1943) 1961. *Pain: Its Modes and Functions.* Chicago: University of Chicago Press.

Carruthers, Peter. 1992. *The Animals Issue.* Cambridge: Cambridge University Press.

Clough, G. 1982. "Environmental Effects on Animals Used in Biomedical
Research." *Biological Reviews* 57: 487ff.

Cohen, Carl. 1986. "The Case for the Use of Animals in Biomedical Research." *New England Journal of Medicine* 315: 14.

Darwin, Charles. (1881) 1882. *The Formation of Vegetable Mould through the
Action of Worms, with Observations on Their Habits.* London: Murray.

———. (1871) 1890. *The Descent of Man, and Selection in Relation to Sex.*
2nd ed. London: Murray.

———. (1872) 1969. *The Expression of the Emotions in Man and Animals.* New York: Greenwood.

Dawkins, Marian Stamp. 1980. *Animal Suffering: The Science of Animal Welfare.* London: Chapman and Hall.

Dean, C. 2005. "Scientific Savvy? In the U.S., Not Much." *New York Times,* August 30.

Duncan, I. J. H., and B. Rollin. 2012. In *What's on Your Plate? The Hidden Costs of Industrial Animal Agriculture in Canada,* ed. World Society for the Protection of Animals. Toronto: WSPA Canada.

Forman, Paul. 1971. "Weimar Culture, Causality, and Quantum Theory: Adaptation by German Physicists and Mathematicians to a Hostile Intellectual Environment." In *Historical Studies in the Physical Sciences,* ed. Russell McCormmach. Philadelphia: University of Pennsylvania Press.

Fox, M. A. 1986. *The Case for Animal Experimentation.* Berkeley: University of California Press.

Frey, R. G. 1980. *Interests and Rights: The Case against Animals.* Oxford: Clarendon.

Griffin, Donald. 1976. *The Question of Animal Awareness.* Los Altos, CA: William Kaufmann.

Hebb, David. 1946. "Emotion in Man and Animal." *Psychology Review* 53.

Hume, David. 1961. *A Treatise of Human Nature.* Ed. L. A. Selby-Bigge. Oxford: Oxford University Press.

Kant, Immanuel. 1984. *Foundations of the Metaphysics of Morals.* Indianapolis: Bobbs-Merrill.

Kelley, W. N., and J. B. Wyngaarden. 1983. "Clinical Syndromes Associated with Hypoxanthine-Guanine Phosphoribosyltransferase Deficiency." Chap. 51 in *The Metabolic Basis of Inherited Disease,* 5th ed., ed. J. B. Stanbury et al. New York: McGraw-Hill.

Kilgour, Ronald, and Clive Dalton. 1984. *Livestock Behaviour: A Practical Guide.* Boulder, CO: Westview.

Kirk, G. S., and J. E. Raven. 1957. "The Atomists." Chap. 17 in *The Presocratic Philosophers: A Critical History with a Selection of Texts.* Cambridge: Cambridge University Press.

Kitchell, Ralph, and Michael Guinan. 1990. "The Nature of Pain in Animals." In *The Experimental Animal in Biomedical Research,* vol. 1, ed. B. E. Rollin and M. L. Kesel. Boca Raton, FL: CRC.

Knight, Andrew. 2012. *The Cost and Benefits of Animal Experiments.* London: Palgrave-Macmillan.

LeBaron, Charles. 1981. *Gentle Vengeance: An Account of the First Years at Harvard Medical School.* New York: Richard Marek.

Lewis, D. B., et al. 1993. "Osteoporosis Induced in Mice by Overproduction of Interleukin-4." *Proceedings of the National Academy of Sciences* 90, no. 24.

Lundeen, T. 2008. "Poultry Missing Genetic Diversity." *Feedstuffs,* December 1, 11.

Markowitz, Hal, and Scott Line. 1990. "The Need for Responsive Environments." In *The Experimental Animal in Biomedical Research,* vol. 1., ed. B. E. Rollin and M. L. Kesel. Boca Raton, FL: CRC.

Mason, John. 1971. "A Re-evaluation of the Concept of 'Non-specificity' in Stress Theory." *Journal of Psychiatric Research* 8.

McNeish, J. D., et al. 1988. "Legless, a Novel Mutation Found in PHT 1-1 Transgenic Mice." *Science* 241.

Michigan State University. 1989. *State News,* February 27.

Mill, John Stuart. 1902. *Principles of Political Economy.* New York: Longmans.

Morton, David, and P. H. M. Griffiths. 1985. "Guidelines on the Recognition of Pain, Distress, and Discomfort in Experimental Animals and a Hypothesis for Assessment." *Veterinary Record,* April 20.

Pew Commission on Industrial Farm Animal Production. 2008. Final Report: "Putting Meat on the Table: Industrial Farm Animal Production in America." www.ncifap.org.

Plato. 1941. *The Republic of Plato.* Trans. with introduction and notes by F. M. Cornford. Oxford: Clarendon.

———. 1961. *Protagoras and Meno.* Trans. W. K. C. Guthrie. Baltimore: Penguin.

Pursel, V., et al. 1989. "Genetic Engineering of Livestock." *Science* 244.

Rawls, John. 1999. *A Theory of Justice.* Cambridge, MA: Belknap.

"Reading Their Minds." 2015. Unsigned review of *Beyond Words: What Animals Think and Feel,* by Carl Safina. *Economist,* July 18.

Rollin, Bernard E. 1989. *The Unheeded Cry: Animal Consciousness, Animal Pain, and Science.* Oxford: Oxford University Press.

———. 1995. *The Frankenstein Syndrome.* New York: Cambridge University Press.

——. (1982) 2006. *Animal Rights and Human Morality.* Buffalo: Prometheus Books.

——. 2007. *Science and Ethics.* New York: Cambridge University Press.

Romanes, George John. 1882. *Animal Intelligence.* London: Kegan Paul.

——. 1883. *Mental Evolution in Animals.* London: Kegan Paul.

Sachs, J. 2008. "The American Anti-intellectual Threat." *Business World,* September 25.

Safina, Carl. 2015. *Beyond Words: What Animals Think and Feel.* New York: Henry Holt.

Sapontzis, Steven. 1987. *Morals, Reason, and Animals.* Philadelphia: Temple University Press.

Schiller, Claire H., ed. 1957. *Instinctive Behavior.* New York: International Universities Press.

Singer, Peter. 1975. *Animal Liberation.* New York: New York Review of Books.

Stout, J. T., and C. T. Caskey. 1989. "Hypoxanthine Phosphoribosyltransferase Deficiency—The Lesch-Nyhan Syndrome." Chap. 38 in *The Metabolic Basis of Inherited Disease,* vol. 1, ed. C. R. Scriver et al. New York: McGraw-Hill.

Stull, C. L., M. A. Payne, S. L. Berry, and P. J. Hullinger. 2002. "Evaluation of the Scientific Justification of Tail Docking in Dairy Cattle." *Journal of the American Veterinary Medical Association* 220: 1298–1303.

Taylor, R. E. 1984. *Beef Production and the Beef Industry: A Beef Producer's Perspective.* Minneapolis: Burgess.

United States Department of Agriculture–National Agricultural Statistics Service. 2015. "Milk Production and Milk Cows." http://www.nass .usda.gov/Charts_and_Maps/Milk_Production_and_Milk_Cows/in dex.php.

Weiss, Jay. 1972. "Psychological Factors in Stress and Disease." *Scientific American* 226 (March).

Wemelsfelder, Francoise. 1985. "Animal Boredom: Is a Scientific Study of the Subjective Experiences of Animals Possible?" In *Advances in Animal Welfare Science 1984–1985,* ed. M. W. Fox and L. D. Mickley. The Hague: Martinus Nijhoff.

Wittgenstein, Ludwig. 1965. "Lecture on Ethics." *Philosophical Review* 74: 3–12.

Wood-Gush, David, and Alex Stolba. 1981. "Behaviour of Pigs and the Design of a New Housing System." *Applied Animal Ethology* 8.

Index